The Face of a Naked Lady

THE FACE
of a
NAKED LADY

An Omaha Family Mystery

Michael Rips

HOUGHTON MIFFLIN COMPANY

Boston · New York

2005

Visit our Web site: www.houghtonmifflinbooks.com.

Library of Congress Cataloging-in-Publication Data
Rips, Michael, date.
The face of a naked lady : an Omaha family mystery / Michael Rips.
p. cm.
ISBN 0-618-27352-2
1. Rips, Michael, 1954– 2. Rips, Nic. 3. Fathers and sons —
Nebraska — Omaha — Biography. 4. Jews — Nebraska —
Omaha — Biography. 5. Omaha (Neb.) — Biography. I. Title.
CT275.R61155A3 2005
978.2'25403'092 — dc22   2004054052

Book design by Melissa Lotfy

Printed in the United States of America

MP 10 9 8 7 6 5 4 3 2 1

FOR NIC, NICK, AND NEILA

The Face of a Naked Lady

THE WOMAN
IN THE BASEMENT

I.

A WOMAN set a coffee before me, and I thought of the first time that I saw a woman fly.

2.

A MONG quiet neighbors, we were the quietest. Father came home every day at the same time, greeted my mother, settled on the couch, and slept; occasionally he would sit in a chair. In either case he would sleep.

At six-thirty he would be called to dinner. After dinner he would return to the couch. Mother would sit next to him. When he finished reading, he would go to his room and sleep.

My father was the appreciative product of his own privileged life. Born in Nebraska, he was Republican, affluent, and content.

As to his relationship with my mother, I heard not a single argument between them. They were respectful and admiring.

3.

MOTHER was sitting on the steps in the hall. In front of her was a box of letters. She pointed to a room in the back. There in tandem on the bureau were his belongings. This was the purpose of my return — to remove what I cared to have. My father had died several years before, and Mother was moving.

But the objects in that room gave off no trace of my father. He fit so smoothly into the order of things, the *circuitus spiritualis,* that he had passed on nothing that was not more perfectly expressed by something nearby; if he had an emotion or thought that was individual to him, it lacked the power of emanation.

I gathered the few things of his and my own that I had decided to take back to New York. Needing a box, my wife, Sheila, and I went into the basement.

After a few minutes, I found a small container and then retraced my steps.

Sheila asked me about a black portfolio that had been slipped behind a cabinet. She pulled it out and laid it on the floor; the portfolio was held together with black ribbons.

An arm, a leg, a torso, another arm, a torso, a head came out of the portfolio. A naked black woman.

Sheets and sheets of a naked black woman, and below each the initials of my father.

On the other side of the basement wall was a small room used to develop black-and-white photographs. Scribbled on the wall of that room was this:

"They do not sweat and whine about their condition,
They do not lie awake in the dark and weep for their sin,
They do not make me sick discussing their duty to God,
Not one is dissatisfied. . . ." Whitman.

.　.　.

Mother was preparing dinner. For as long as I could remember my family had a cook. The ablest was Mary. But even the worst were capable of being taught, and my mother did a very good job of that. They were different from the meals I would get at our neighbors'.

Claire was one of our neighbors and I enjoyed visiting her. One evening at Claire's, we heard her brother, Ronald, singing "Surrey with the Fringe on Top" from the musical *Oklahoma!* That was unusual because Ronald had for years sat quietly in his room. I imagined that he was writing or composing or juggling and that one day I would hear that he had won a prize.

Claire went straight to his room. She wanted to share in his happiness.

What she saw was a happy Ronald lying on his back rotating a live chicken on his manliness, singing "Chicks and ducks and geese better scurry when I take you out in the surrey. . . ."

We made our way back to Claire's room. Minutes later Ronald passed the door.

He was upset.

Recounting her last minutes, he explained that in enjoying a chicken the greatest pleasure comes when the chicken's neck is broken, causing a "death shiver" that Ronald found "impossible to duplicate"—suggesting that Ronald had only settled on chickens after experimenting with other animals.

Several days later, I found myself back at Claire's. It was the late afternoon and Ronald's mother had returned to the house with a friend who was visiting Omaha from the East Coast. Claire's mother invited me to join the family for dinner.

When Claire and I came to the table, everyone was there but Ronald. He was still in his room.

Claire's mother brought out a delicious first course. Before we had finished it, Ronald arrived. His mood was good.

Having cleared the table, Claire's mother returned from the kitchen with the tetrazzini. Then I saw it. Ronald's face rippled. There was only one conclusion: we were about to eat his lover.

As I reflected on this, the woman from the East Coast, who was sitting to my left, placed a good-sized portion of tetrazzini on her plate. Believing that one is obligated to warn one's dinner companion that she is about to consume a dish that has been inseminated by another guest at the table, I leaned toward the lady from the East Coast and whispered, "The chicken was murdered."

There was no response.

With Ronald's lover now inside her mouth, I bent down, pretending to have dropped my napkin, and turning my head upwards from next to her knee, whispered, "There's semen in the chicken."

That did it.

In retrospect it has occurred to me that I'd simply substituted an obvious observation with an obvious and repulsive observation, and the woman's inability to finish her meal had less to do with the chicken than with me. Ronald had succeeded in making me more revolting than Ronald himself.

I would like to say that Ronald was now in the musical theater, but the truth is that I do not know what happened to Ronald.

We left Omaha on a Sunday. On the way to the airport, we passed the Civic Auditorium. It is where I saw the woman (Miss Rietta) fly.

4.

IN NEW YORK, I went back to sitting in a coffee shop on Ninth Avenue.

A blond woman approached my table. She was from the seminary. She was a friend of the Bearded Priest and asked me if I'd seen him. She was concerned.

Before becoming the Bearded Priest, he had raised bird dogs and before that worked in a lumberyard. As an Episcopalian priest, he had lived among the spiritists of Haiti and now spent his time reading Emmanuel Levinas and fishing off a pier on Fourteenth Street. I had taken note of him because of his resemblance to Whitman and because of Levinas.

A Lithuanian Jew who had moved to Germany in the late 1920s, Levinas attended Husserl's last semester of teaching and Heidegger's first. Levinas's translation of Husserl's *Cartesian Meditations* brought phenomenology to the French and specifically to Sartre. During the war, Levinas was taken prisoner by the Germans and placed in a concentration camp.

Levinas survived the war. His family did not. It was at this time that Levinas was introduced to the Talmud, and it was to the Talmud that he devoted much of his writing.

I told the woman that I'd not seen the Bearded Priest.

She sat down. She was attractively built, her face was a deep pale, her eyes soft and alarmed. We began to speak. She told me that she had been brought to the priesthood by accident, a chance reading of the story of Saul.

My father had told me this story.

He was not known for talking. Even to his sons he spoke little. But this story he had told me, this story among a handful of others, randomly offered and separated by long stretches of time. Some of these stories I remembered, some I did not. Some of my father I remembered; some I did not.

I asked the woman from the seminary to retell the story. She did and as she did the late afternoon light withdrew and all was quiet.

The story of Saul is found in the first book of Samuel.

It begins with the elders of Israel approaching the aging prophet Samuel with the request that he name a king. The Israelites, weak and under siege, were concerned that when Samuel died, they would be left without a leader. Samuel put the request to the Lord.

The Lord warned the Israelites against a king.

The Israelites insisted.

Granting their request, the Lord chose Saul, a modest young man from the smallest clan of the smallest tribe.

Saul immediately proved himself, leading the Israelites to victory over their enemies. Before his battle with the Amalekites, Saul had received a specific instruction from God: the Amalekites were to be "entirely destroyed," including the women and children and cattle.

But Saul did not listen to the Lord. Having defeated the Amalekites, he spared the life of the Amalekite king, Agag, and the best of the Amalekites' animals.

When Samuel discovered what Saul had done, Samuel scolded Saul. Saul returned that Agag was in the custody of the Israelites and that the animals were to be used as sacrificial offerings to God. Samuel cut Saul short: "Does the Lord delight in burnt offerings and sacrifices as much as in obeying the voice of the Lord?" Leaving no doubt as to where he stood on the matter, Samuel had Agag dragged out and executed.

Displeased with Saul, God sent an Evil Spirit to torment

him. At the same time, God instructed Samuel to summon David, whom God had selected as the next king of Israel. Jealous of David ("Saul has his thousand, David his ten thousand") and tormented by the Evil Spirit, Saul attempted to slay David. While David was in bed with his wife, Saul's men entered David's house. David's wife urged David to escape, replacing David in the bed with a teraph topped with goat's hair. The substitution allowed David the time to flee.

The story ends with Saul leading the Israelites in battle against the Philistines. As the battle neared, Saul, desperate to know whether the Israelites would prevail, found himself cut off from the voices that could assist him: Samuel was dead, Saul's dreams revealed nothing, and there was a standing interdiction (issued by Saul himself) against seers. Defying his own decree, he approached the Witch of Endor.

Saul's plan was to have the Witch find Samuel and bring him back to the living. Saul would then consult Samuel on the battle with the Philistines.

When the Witch recognized Saul, Saul promised the Witch that she would not be punished for violating the law against necromancy. So assured, the Witch retrieved Samuel from the dead.

Samuel began by chastising Saul for past offenses. As to the battle with the Philistines, Samuel informed Saul that neither he nor his sons would survive. The next day, Saul, surrounded by the Philistines, impaled himself on his sword. The Philistines removed Saul's head.

At the point in the story when Saul approached the Witch of Endor, the voice of the woman in the café was no longer the voice of the woman in the café. My father was finishing the story and by the time he had finished the story, I knew that I would be leaving New York and returning to Omaha. I would find the black woman in the picture and through her find my father. I would retrieve him from the dead.

5.

IN OMAHA, I phoned detectives. After talking to three or four, I set up an appointment with Frank Williams. Williams, according to the other detectives, had ways of getting information that no one else had.

To meet Williams, I traversed the part of Omaha that had been built since the war — the shopping malls, the strip malls, the stores the size of malls, the car lots, the fast-food restaurants, all close to the earth, close to its color — emerging upon that neighborhood that is older, blacker, and lighter. I stood, finally, outside a locked bar. From a surveillance camera mounted near the door, Williams watched me.

The door opened.

The interior was dark. The door was locked behind me.

A black man led me to a stool in the empty bar. He poured me a drink.

I explained why I was there.

Williams asked me to follow him. He took me to a place where we could speak without being overheard.

In the basement of the bar was a room with three tables. Passing through that room, Williams removed a set of keys and unlocked a door. On the other side of the door were a desk and a series of monitors.

After reviewing the monitors, all of which were trained on the empty parking lot, he questioned me.

"Have you talked to anyone about this?"

A few.

"What do they know?"

Nothing.

"Did your father have a best friend?"

Yes.

"Have you talked to him?"

"No. He is sick and recognizes no one, including himself. He's in no position to help."

"I'll be the judge of that."

Williams checked the surveillance monitors.

"He's completely unconscious," I insisted.

"Not so fast!"

"Not so fast?"

"Have you heard of the brain machine?"

"No."

"I've used it to solve many cases."

"A brain machine?"

"The machine is in Iowa. It goes into your brain and reads thoughts you didn't know were there. You can't lie to it, can't fool it; and once you've answered it, you can't undo it."

Williams checked his watch.

"Come back at three and don't eat."

When I returned to the bar, I picked up Reggie, a friend who had experience with detectives.

Williams had been to a funeral and was in a good mood.

"Wait for me downstairs."

Reggie and I sat at one of the tables outside Williams's office.

Williams appeared at our table.

As he sat down, a door opened at the other end of the room and a naked woman walked out. She was followed by another woman. Also naked. And then another. There were six women in all—all black and all naked.

After a brief discussion among the women, one began to

move around the room. She was five foot five or six and weighed three hundred pounds.

At the center of the room, she stood still for several minutes as if readying herself for an oration. Then, suddenly, she flipped herself upside down. As she balanced on her head, two men strode from the shadows. One took a position on her left, the other on her right. Bending forward, they stared fixedly into the fullness of her womanhood.

Williams, who had disappeared, arrived at our table with a platter of food.

We were joined by the naked women. We all began to eat.

Perhaps this was Williams's idea of a lineup, not having realized that the woman I was looking for would be in her seventies. On the other hand, he may have thought that I was the one looking for a black woman and that the story about my father was a coded request. Or this may just be what Williams did in the afternoon.

As we drove from the bar, Reggie offered this: "Williams knows the woman. He's got upside-down women, he's got brain machines and trays of food—everything but the person you're looking for, and the reason he doesn't have her and doesn't ask you any questions about her is that he knows who she is. Williams knows everything about your dad and his woman and he's protecting them."

"And what," I asked, "do you think he knows?"

"Easy. Your dad was taking the women from his factory to his office and painting them; they went along with it because he was paying for it or he was having sex with them or he was paying them and having sex. He got the factory from his folks and decided to fill it with colored women who looked good with their clothes off. It was part plantation, part art studio."

"A demented Schindler?"

"They go along with it because he's the boss or because they like him or feel sorry for him. Finally he falls in love with one of them. She becomes his woman, and he can't bear the idea of throwing out her portrait."

If Reggie was right and my father was painting his mistress, then he was a man that I did not know, for I had never seen him with another woman, never heard rumor of such, and from everything that I knew he was devoted, throughout his marriage, to my mother.

But if the woman in the painting was his mistress and was still out there, I would find her. And when I found her, I would find her children, my half brothers and sisters. There would be the distasteful surprise, denials, hostility, but soon all would turn — awakening, reconciliation, friendships would follow. I readied myself.

Down the street from the hotel where I lived in New York was a large advertising agency. That agency was founded in Omaha, though its principal office was in New York. The fellow who started the agency had been a friend of my family's.

There was a bar between the agency and the hotel, and people from the agency were often there after work. It made sense that before I began my search I would meet the man in the fine suit.

The man in the fine suit was born in Los Angeles, had gone to school on the East Coast, and had worked for years in New Orleans. He now lived in New York. For ten years, between New Orleans and New York, he resided in Italy, traveled through much of Eastern Europe, and took his vacations in Montenegro. As to clothes, he preferred a tailor in London.

The mention of London recalled a remark that I'd heard years before — that Omaha was the ideal place in which to test-market products. I had not known this, and I asked the man in the suit what it meant. The man in the fine suit said that people in his business understood that America was white, docile, and middle class and that Omaha had taken on the reputation, rightly or wrongly, as that city which was most thoroughly white, docile, and middle class. What the man in the suit found so laughable, so disappointing, about the Midwest was that the Midwesterner, once the symbol of the individual—the ones who had stayed behind when the depression of the 1890s had driven the weaker of them west—had converted so quickly into its opposite, all within a generation and all without evident upset.

I had not known this. We are often in greatest ignorance of the place to which we belong.

I was prepared to concede the truth of what he had said save for one thing—the black woman in the basement; more precisely, my father painting that woman. By all accounts he was the most vacant of Midwestern men, and yet there he was, paintbrush in hand.

HIS NAME

I.

No one in my family knew anything about the black woman, and a month after my visit home I had still not heard from Williams.

At this point the brain machine came to mind. I'd located the brain machine and its inventor, but they were in eastern Iowa and the brain machine was far too large to be transported to Omaha. This meant that I would have to get my family and my father's best friend into a bus and drive them hundreds of miles, where they would sit in a barn, attached to a gigantic device, answering questions about my father's life. All on the hope that somewhere inside them was the naked woman who knew my father.

But there was another way of solving the mystery of who this woman was and what relationship she had to my father. As the brain machine scanned brains, I would scan his life, following the accretion of his thoughts and moods from the time he was a boy until the moment he picked up the paintbrush. At that point everything would be clear to me, including the name of the woman. With that, I would track her down and question her as to my father and the life that he had kept from his family.

The Slosburgs were the place to start. Bud Slosburg, his wife, Eleanor, their sons, Dickey, David, and Danny, and their daughter, Mary Jo, lived around the corner from us. They had straight, rust-colored hair and light eyes; all were athletic. They had a pool and a tennis court, and all of them had fun. Not pretending to have fun; they actually had fun. One may not think of Jews when one thinks of Nebraska, but I do, and when I think of Jews in Nebraska, I think of the Slosburgs.

Bud Slosburg's brother is Bussy Slosburg, and Bussy Slosburg was a classmate of my father's at Dundee Elementary School. They grew up together in Omaha, and Bussy's children —Jack and Jill—were friends of my brothers. Bussy may be the person with the earliest memories of my father.

Today Bussy lives in a suburban home in the western part of Omaha—a house that has a modest front lawn and a small pond that Bussy and his wife, Marianne, share with others.

As I sat with Bussy in the living room of his house, he explained that when my father transferred into Dundee Elementary School, he knew no one. Prior to this he had been living in South Omaha, where his parents owned an optical business. South Omaha had been a separate city until just five years before my father was born, when it was incorporated into Omaha. With their business doing well, my father's parents decided to move to the neighborhood of Dundee, one of the most affluent parts of the city.

"At first your father had no friends," Bussy responded. "The other boys were not interested in having him around, and your father showed no interest in being part of our circle—he was a ten-year-old loner." Bussy stopped and gazed outside. A bird had landed in the backyard. He watched it and then resumed: "Your father was not much of a student. He didn't do much and didn't want to do much. I don't think he was in any of the clubs."

Bussy, by contrast, had quite a nice record of club memberships.

"By the time your father entered Central High School, he had fallen in with the Milders. That was his gang—Tono, Pete, and the others."

As Bussy spoke, Marianne entered the room and took her place next to her husband. In between Marianne and Bussy was a painting by their daughter, Jill. She painted but also made jewelry. My father had bought one of her necklaces for my mother.

When we finished talking, Marianne asked me if I would like to look around the house. As I was about to get up something occurred to Bussy.

"Something has occurred to me," Bussy said. "Your dad was the first to drink."

"Yes," added Marianne.

I'd heard this before. There were stories of the Ripses, Slosburgs, Shermans, and others drinking away at the steak houses, bars, and clubs of Omaha—the Sparetime, the Colony Club, the Cave Under the Hill, the Blackstone Hotel, and Chez Capri.

Across from Jill's painting was a shelf of books. "I have a special interest in female novelists from Britain," Bussy offered.

I got up from my chair.

Bussy again stopped me.

"Do you know why your dad's name was Nick?"

I did not know. His given name was Norman, but no one

called him that. It was always Nick. I was disinclined to think about nicknames, for they had always confounded me. There were a great number of them among my parents' friends (Bussy, Trickle, Pussy, Penny, Ducky, Doc), and if they were not a semiconscious attempt to assume control over their neighbors, then they were a primitive device to escape the inevitable to which each of us is chained—a mask, a simulacrum. The summoners might be fooled and take Pussy and not Margaret or whatever Pussy's real name was.

Despite my aversion to nicknames, I had a special interest in the question that Bussy posed. My daughter, Nicolaia, was named after my father.

Bussy proudly provided the answer to his own question.

"I was the one who gave him the name Nick, because he was the first of us to smoke cigarettes. I used to call him Nicotine, and after a few years Nicotine Rips got shortened to Nick, and that was what we called him from then on."

A fine story, though it put me in the position of being one of the few people who had named his child after a carcinogen.

"And one more thing."

I turned to him.

"I may be wrong—Marianne, correct me if I'm wrong—but Nick was the first to get laid."

Marianne nodded.

The bird that had been in the yard was now at the edge of the pond, staring into the water. The bird did not go in, knowing that its waves would disrupt the color. A dog came to see the bird but knew not to look its way. Only man ventures so close that it destroys that which it seeks to find.

2.

B USSY was right about my father falling in with the Milders. According to all accounts, he spent a lot of his time with them. As my mother tells it, he was especially engaged by the figure of Morris Milder. My mother is smart and extremely discreet: for her to make any statement about my father suggested significance.

What my mother did not say was that my father's involvement with the Milders did not end with his childhood or with the death of Morris Milder. Morris Milder had one sister, Anne, and my father would marry her daughter. Morris would become my great-uncle.

To understand why the very young Nick Rips would be fixated by the much older Morris Milder, I needed to speak with Mike Milder. Mike, who is now in his seventies, was the son of Morris's brother, Hymie.

When I was growing up, Mike and his wife, Ducky, lived in a pagoda surrounded by a moat. During the sixties and seventies, riots broke out in Omaha. At the time, Mike Milder was one of the leaders of the Republican Party. However slanderous the attacks on his Party, Mike remained good-natured and open to conversation, many of which he held in his pagoda.

I drove out to Mike and Ducky's. The city had long since seized the pagoda and moat for public use. Mike and Ducky now lived in an understated and attractive area, not unlike that in which the Slosburgs lived.

Sitting in Mike's living room, the walls covered in oriental screens, Mike explained that at the time my father first became acquainted with the Milders, they were one of the most promi-

nent families in the city and that this was in great part the doing of Morris Milder.

If my mother had led me to Morris Milder, Morris Milder
led me to Tom Dennison (the political boss who controlled
much of the city in the 1920s and '30s), for it was Morris's as-
sociation with Tom Dennison that made Morris one of the
wealthiest and influential men in the city. In the year 1919,
Morris Milder, according to the newspapers, was worth well
over a million dollars. While Morris had his own businesses,
there is little question that he received considerable income
from prostitution, bootlegging, and gambling—activities that
he and Dennison controlled.

Dennison was tall, rugged, and had a cauliflower ear.
Known to favor dynamite as a means of relieving himself of his
enemies, his nickname was Pick Handle Tom, from an incident
in which, confronted by an armed attacker and having forgotten his dynamite, he seized a nearby pick handle and pum-
meled the man—pick handles being relatively easy to come by
in Omaha. And yet Dennison was soft-spoken—his hushed
voice concealed a slight stammer—and well dressed. Before
coming to Omaha, he had been involved in illegal gambling, a
business interest he brought with him to his new home. From
gambling, Dennison took positions in saloons and other illegal
enterprises, making certain that he paid off or strong-armed
those politicians who could provide him with protection.

The mayor at the time was Cowboy Jim Dahlman. Raised in
Texas, Dahlman left the state after killing a man. He traveled
under an assumed identity until he was acquitted of the crime
by a judge who ruled that the killing was in self-defense. Instead of returning to Texas, he stayed in Nebraska, where he
became mayor of Chadron and then Omaha. Dahlman, who
wore a cowboy hat, suit, and a bow tie, was fond of randomly

roping people on the street. As to gambling and prostitution, Dahlman had an extremely open mind. As to drinking, Dahlman was strongly for it, having broken with his friend William Jennings Bryan, who was strongly against it.

Dahlman was not about to interfere in Dennison and Morris Milder's affairs. It is said that in the entire time that he served as mayor, covering over two decades, Dahlman never once mentioned Dennison's name. Help for Dennison also came in the form of the chief of detectives, a Jew by the name of Ben Danbaum.

With the politicians under his control, Dennison was approached by men from all parts of Omaha society to help them secure contracts, easements, deposits of public funds, and immunity from government inspectors. For the jobs he provided working men and women in Omaha and the loyalty he showed them, Dennison was widely liked.

My father would have been familiar with Dennison because of an incident that occurred around the corner from the Miller Hotel—a hotel in South Omaha owned by my father's grandparents. The incident involved a bootlegger from South Omaha by the name of George Kubik.

In November 1931, George Kubik and his son, also a bootlegger, attended the fight between Jack Dempsey and Bear Cat Wright. After the fight, father and son picked up their morning delivery of alcohol. Returning to his home in South Omaha, down the street from the Miller Hotel, George Kubik was abducted by three men who drove him to a location in the western part of Omaha, where they fired several bullets into his body and head. With their job finished, they left.

But Kubik was not finished. He crawled to a nearby home, knocked on the door, introduced himself, and then reported the details of his own assassination. Having finished with the story, he died. Though Dennison was never tied to the murder, it was suspected that Kubik was buying and distributing liquor without clearing it with Dennison; if Dennison did not order his death, others did, with Dennison's approval.

When my father was eleven years old, he made a point of forgoing his friends in the playground to attend the wake of George Kubik—one of the few details of my father's early life that he cared to mention.

The death of Kubik, prominently featured in the Omaha papers, was soon followed by another murder—this one of even greater notoriety. Two days before Christmas of 1933, Harry Lapidus, a prominent businessman, philanthropist, and leader of the Jewish community, was gunned down as he pulled to a stop near Hanscom Park. Those gunmen were more efficient than the ones who had shot Kubik, putting all of their bullets into Lapidus's head—one into his face, one into his ear, and the third into the top of his skull.

Lapidus had been involved in efforts to put an end to organized crime. Moreover, Lapidus's son-in-law was an assistant attorney general who was well known for his efforts to end corruption in Omaha. Dennison had motive to kill Lapidus and was therefore the focus of the prosecutor's investigation.

If Dennison believed that he was too powerful to be indicted, he did not anticipate the backlash that resulted from Lapidus's death. Reformers committed to ending Dennison's control over the city succeeded in obtaining a federal indictment against Dennison and fifty-eight members of his organization. The trial, which lasted for two months, ended in a hung jury, but the testimony against Dennison put an end to his rule.

After the trial, Dennison spent increasing amounts of time in California. Two years before his trial had begun, the seventy-two-year-old Dennison, a widower, married a seventeen-year-old Omahan by the name of Navajo Truman. A year after the trial, she filed for divorce. A year after the divorce, Dennison died in California of a hemorrhage following a car accident. He was buried in Omaha at Forest Lawn Cemetery. Over a thousand people attended his funeral.

3.

SITTING in Mike's living room, I was told one last story about the world of Morris Milder in the 1920s and 1930s — the world of prostitution, bootlegging, and gambling that had intrigued my father as a young man. To put it this way, though, misrepresents the way in which my father was seeing these events, for as it turns out he was not the child staring at something adult and unknown through a partially shaded window. He already knew that world; he had already been inside.

Mike's story began with Morris approaching his brother, Hymie, to ask a favor. Morris needed someone to deliver a document to a friend of Morris's in Chicago. Hymie was a young man and happy to do whatever he could to help his older brother.

A couple days later, Hymie was on a train to Chicago, carrying with him a folder which he was to open once he got there; in the folder was an envelope and on the envelope was a name and an address — Hymie was to hand the letter to the person whose name was written on the envelope. The rest of the time in Chicago was his own.

After Hymie checked into the hotel, he tucked the folder under his arm and walked out of the lobby and onto the street. Opening the folder, he found, as his brother had promised, the address to which he would deliver the letter. It was around the corner from the hotel.

As to the name on the envelope, it could not have been clearer: Al Capone.

Hymie made his way to the building where Capone kept his office, and when he told the guards in the lobby who he was and why he had come, they let him up. They were expecting him. Arriving at Capone's office, Hymie again reported who he was and why he had come.

Hymie waited.

A door opened.

The man who walked through the door introduced himself as Ralph Capone, Al Capone's brother. Hymie turned over the letter and Ralph disappeared through the door. It was some time before Ralph returned. When he did, he said to Hymie, "Tell your brother that Al says it's okay. Any friend of Morris is a friend of Al's."

The transaction in question had to do with a man in Omaha who wanted to set up a speakeasy. Since Capone provided the liquor for Omaha, the man would have to clear his speakeasy with Capone. That's where Morris came in: the man knew that Morris was a friend of Capone's and that if the request came through Morris, there was a good chance that Capone would approve it.

Hymie, pleased that everything had gone so well, thanked Ralph and left the building. As he reached the street, a man stopped him and said that he had been sent by Al Capone to give Hymie a lift back to the hotel. The man's car was parked in front of the building. Hymie said that he appreciated the offer but had decided to walk back to the hotel.

The man explained to Hymie that Al Capone's instruction was to give Hymie a ride to the hotel and that Hymie was probably better off taking the ride. Hymie got into the car.

The car drove a couple blocks, turned the corner, and then parked next to the hotel.

Hymie thanked the man and got out of the car. As Hymie entered the hotel, the man from the car was standing next to him. This time he was carrying a satchel. "Al Capone," the man explained, "wants to make certain that you get to your room safely." At this point, Hymie grew nervous.

With the man following him, Hymie entered the lobby of the hotel. In the center of the lobby, where he was visible to everyone, Hymie turned and thanked the man for his kindness. But Al Capone wanted to make certain that Hymie was accompanied to his room, and Hymie, who no longer believed that the man was necessary for his safety (he had come to the opposite conclusion), had no choice but to proceed to his room with the man behind him. Al Capone had obviously taken offense at Morris's request, and Morris, having anticipated this possibility, had sent his sweet but expendable younger brother in his place.

Standing in the hallway in front of his room, Hymie made one last attempt to convince the man to leave him alone. The man responded by pushing Hymie into the room and instructing him to take a seat on the bed. The man pulled a table in front of the door and told Hymie to close his eyes. It would be over in a few seconds.

Sitting on the edge of the bed, Hymie heard the man reach into the satchel. Hymie began to weep. There was a noise and, as the man had said, Hymie felt nothing.

Hymie felt nothing because no bullet had entered his head. He waited for the second shot. The shot did not come. Slowly Hymie opened his eyes. The room was empty. The noise he had heard was the sound of the door closing. In front of the door was the table, and on the table were four bottles — bottles of fine Scotch — with a note: "Enjoy yourself in Chicago. Al Capone."

Morris Milder's power in Omaha, though considerable, was shared with Dennison, and when Morris and Dennison had a falling out, Morris left Omaha. That Dennison's affection had changed was signaled when he firebombed Morris's house.

Following the bombing, Morris moved his family to St. Joseph, Missouri, where, protected by Tom Pendergast, the mob boss in Kansas City, Morris was beyond Dennison's jurisdiction.

4.

MORRIS'S brother, Hymie, my great-uncle, remained in Omaha and became a successful businessman. He owned an oil company and, like his brother, carried a great deal of weight with local judges and politicians. One of the last times I saw Hymie was when my brother Harlan and I were visiting a shopping mall in west Omaha. Harlan had come home from college and the two of us decided to take a ride. The mall was crowded and we were forced to park near the back of the parking lot. As we approached the mall, we noticed an ancient black limousine parked on the sidewalk. There was only one person in Omaha who drove such a car: my great-uncle Hymie.

As we stood next to the car, the window dropped. Smoke rolled from the interior. He was smoking a cigar; his pants stretched up and around his belly, tickling the bottom of his nipples; his eyes were framed by black square glasses.

"Gentlemen," he said, "I'm getting old and I've one piece of advice to give you as you go forth into the world."

We waited. The waves of heat expanded and contracted his face.

"Marry a black woman."

The window closed. Three decades later, I heard his son laugh.

"Frankly, I can't believe he said that; I was the one who taught him that, but I thought he wasn't listening."

"You taught him?" I said.

"Of course. We used to sit next to each other in the office and talk about things and every once in a while I would say to him, 'Dad, you work with blacks, hire blacks, have black friends, it's time you admitted that blacks and whites should have sex.'"

In 1919, when Hymie was a young man, Will Brown was taken into custody for the rape of Agnes Loebeck, who was nineteen years old and white. Brown was black. After Brown's arrest, a group of boys (former classmates of Loebeck's) gathered at Bancroft School at two o'clock in the afternoon and from there marched down to the courthouse where Brown was being held. Convinced that the boys posed no threat, fifty police officers, who were being held in reserve, were sent home.

Around five o'clock, adults joined the boys. The crowd, which quickly numbered four thousand, charged the courthouse. The police attempted to turn them back with water hoses.

The crowd responded by tossing bricks through the windows of the building and looting nearby stores for firearms. Reports at the time recorded the theft of over a thousand guns.

By six o'clock the mob had broken through the police line and made their way into the courthouse. At the same time, those outside the building were beating every black near the courthouse. Whites who came to the assistance of blacks were also beaten.

The police put up their last defense on the fourth floor of the courthouse. Frustrated, the rioters set fire to the building. This forced the police to move Brown and other prisoners to the roof of the building. As the fire rose around them, the prisoners became hysterical. Several tore out their hair, others attempted to throw Brown off the roof.

At this point, Mayor Edward P. Smith presented himself at the front of the burning courthouse. He ordered the crowd to pull back. The crowd refused.

Smith then shouted: "If you must hang somebody, then let it be me."

Taking him up on the suggestion, the mob threw a rope around his neck and dragged him to a lamppost. The police intervened and got Smith into a car, but the car was toppled and Smith was pulled out. At Harney and Sixteenth Street, the rope around Smith's neck was swung over another pole and his body pulled into the air. The lynching of the mayor would have been complete had Ben Danbaum, the chief of detectives, not arrived. Driving his car at full speed through the throng, Danbaum stopped beneath the mayor. Someone from inside the car jumped out and pulled Smith inside. Danbaum drove the car out of the crowd and delivered Smith to Ford Hospital, where he remained for several days in critical condition.

As the mob met resistance on its way up the stairs of the courthouse, some took to scaling the outside of the building so as to reach Brown on the roof; headlights were turned toward the courthouse to assist the climbers. At this point, the mob had grown to twenty thousand.

Almost half a day after the rioting had begun, Will Brown was in the hands of the rioters. Dragged from the courthouse,

Brown was hung from a telephone pole and fired upon by shotguns, rifles, and revolvers. The crowd then pulled Brown down and tied him to the back of a car. His corpse was hauled through the business district of Omaha, after which it was soaked with the oil from lanterns and set aflame. The rioting did not end until federal troops arrived at three in the morning.

Still unknown is the identity of the man who jumped out of Danbaum's car and took the noose off the mayor's neck. Some say it was Russell Norgard, others say A. C. Andersen, and at least one account has it as Morris Milder.

Not many years after the riot of 1919, in the middle of winter, another black man was accused of rape. I had heard of this incident from my father.

With the suspect in the station house, a crowd assembled. Hymie happened to be in the station that night, chatting with Danbaum. Looking outside, they saw the crowd.

Hymie asked Danbaum what he planned to do.

"Nothing," Danbaum replied. "Smith nearly got himself hanged, and I'm not next."

Danbaum's officers felt the same way, for when the crowd tried to get into the police station there was no resistance. In no mood for what was about to happen, Hymie put on his coat and made his way out of the jailhouse as the increasingly excited crowd pushed their way past him.

The crowd, having arrived at the cell, found no prisoner.

Danbaum pleaded ignorance, and the rest of the police force, embarrassed by the lapse, went looking for the suspect. Nobody found him.

Before leaving Mike's house, I asked whether he understood my father's interest as a child in Morris and Hymie Milder, Tom Dennison, and the funeral of George Kubik. Mike told me that he did not know. For that I would need to look elsewhere.

As Mike got up to see me off, I noticed a portrait of Hymie hanging on the wall. It was painted toward the end of his life: he was handsome, with bright pearl skin; but he was always overweight and had been his entire life. People said, for example, that his topcoat was so large that if a man was tucked beneath it, Hymie could walk out of a building and no one would notice.

5.

CENTRAL HIGH SCHOOL, a large neoclassical building, sits on the top of a hill overlooking the Missouri River. The hill was originally occupied by the territorial capital. Shortly after Nebraska entered the Union, the capital was moved to Lincoln, the old capitol building was torn down, and a high school was built in its place.

From the day it opened, Central High School has been filled with the races, classes, and nationalities that had built Nebraska. The school has produced philosophers (Saul Kripke), Nobel Prize winners (Lawrence Klein), and Hall of Fame athletes (Gayle Sayers). There was no question of my father not going to Central High School—his mother had gone there and his friends and relatives would go there.

The relatives presented a small problem. There were three Norman Ripses at Central High School at the same time. The principal, who found no amusement in this, called them into his office and gave them new names. My father, because he was born in July, was to be known as Julian, and from that day on my father became Norman Julian. He would not appear in the principal's office again until many decades later when he came with me in his pajamas.

From Central, looking east, one can see the Missouri River and across the Missouri, Council Bluffs, Iowa.

In the evenings, my father would drive to the sand dunes on the Platte River, a tributary of the Missouri. Usually this involved women and drink. Sometimes he went there to read or just stare into the sky. The dunes on which he lay were the largest in the Western Hemisphere, 20,000 square miles of sand hills shaped by the wind and held in place by grass. Beneath the dunes was the Ogallala Aquifer, the world's largest underground supply of water. With the sand on top and water on the bottom, the ocean had been turned upside down.

Father always had a car so it was Father who would be driving back and forth to the dunes. If he had been drinking and was worried about falling asleep, he would pull into a diner and order a cup of coffee. Norman Lincoln, a classmate and early friend of Father's, recounted that while others would drink from a mug, Father would dip his tie into the coffee, wait until it was saturated, and then, lifting it above his head, drain it into his mouth. On the road his coffee dangled beneath his chin. If he needed a sip, he could suck on the tie, his eyes never leaving the road.

One evening, Father and some of the Milders were at my father's house when someone had the idea of driving to Kansas City for the evening. With his parents (Aron and Esther) away on vacation, he took the keys to their car and off he and his friends went. In those days, Kansas City was a five-hour drive, and once there, they decided to stay a day or two longer.

Aron and Esther returned from vacation: the doors of the house were open, the lights on, the Victrola running, car missing, and no sign of their son. Esther Rips was known for a volcanic temper and when her son finally returned home he got a full blast of it. By all accounts, it singed him not a bit. And that, as his brother Sheldon recounts, was one of the most impressive aspects of my father: an impenetrable calm.

Sheldon, seventeen years younger than my father, tells the story of running into his brother's bedroom one morning to warn him that Esther was coming.

"She's going to kill you," Sheldon shouted at his brother.

Father ignored him.

"The car!"

This caught my father's attention.

The night before, father had driven home, missed a turn, and ended up in a lake down the street from the house. Deciding that the car was likely to still be there in the morning, he walked home and went to sleep. His mother's tantrum was, as Sheldon tells it, received with indifference.

But on most occasions, my father's friends told me, my father had an easier means of avoiding confrontations with his mother. Rather than returning home after a night of drinking, he would drive to South Omaha and check into the Miller Hotel, the hotel owned by his grandparents. He was guaranteed a room, and the girls at the hotel would make sure he got up in the morning. He would have breakfast with his grandparents or walk to Johnny's Café, where they opened early to feed the men who came in on the cattle trains.

My father's disengagement was not limited to his mother.

Norman Lincoln, who was as close to my father as anyone during this time, observed that my father "lived within himself." It was a reflection shared by everyone who knew him.

Norman Lincoln still lives in Omaha with his wife. She was born Margie Lipp and she too went to school with my father. She has a very specific memory of him: "Your father liked to read," she told me. "He would sit alone at a bar up the street, the Dundee Dell, and read books."

Schoolbooks?

"Other books. During high school, whenever I wanted to have an interesting talk, I would go up to the bar and your dad would be sitting there."

She had one other memory of my father: "He liked women."

Mother had provided me with a list of those whom she thought would be helpful in providing information on my father as a child. One of the names on the list was a man whom I knew. I phoned him to have dinner.

We met at a restaurant on the western edge of the city. He had a drink, I had a drink, the waitress brought over the menus. The waitress reappeared to take our orders. He ordered, I ordered, but he stopped me. I was making a mistake. The dish I should be ordering was the scaloppini. He said this with such certainty that I ordered the scaloppini.

As the waitress left, he filled me in on the dish. The recipe was not new to the restaurant. Rather, it was an old recipe that had begun at the Italian Gardens and when that closed was picked up by a chef at a private club. That chef passed away, and the new chef at the club continued to serve it but so altered the original as to nearly ruin it. People started to remark on this, and a couple of men prevailed upon the owner of a new restaurant to put the scaloppini, in its original form, on his menu. He did so, and the popularity of the item guaranteed the success of the new venture. The scaloppini had two or three more homes before it ended up at the restaurant where we were sitting.

The waitress returned. They were out of scaloppini. I ordered something else.

As the waitress brought our coffees and dessert, I asked my

father's friend what I had come to ask: what could he tell me about my father, the man whom he had known for seventy years.

Putting down his fork, he engaged in a prolonged reflection, looking first this way and then that. Finally, he cleared his voice:

"Your father . . . he was okay."

Was there anything else?

"No."

Was there anyone else who might help me?

"I doubt it."

Then he paused. He was looking off. He was about to say it. The answer to the mystery.

"A woman," I said quietly.

There was no response, but his stare was becoming more fixed.

"You are seeing a black woman," I pressed in the tone of a hypnotist.

His expression changed, his eyes now back on me.

"A Negro? I'm seeing a Negro?"

I nodded.

"No Negro!" he snapped.

I turned around. What he was seeing was the dessert cart.

If he had been thinking of ordering dessert, which he undoubtedly was, I had changed his mind. He waved for the check.

6.

ACCORDING to Marvin Taxman, who, along with Norman Lincoln and Bussy Slosburg, still lives in Omaha, my father loved literature. I asked Marvin what my father read and he gave me these: "Halliburton, Wolfe, sometimes Maugham."

He stopped and then finished with this: "Nick and I, having read a couple books, would drive up to the dunes, lie on our backs, and talk about those books."

On July 14, 1936, with her son turning sixteen, Esther Rips, my father's mother, decided to throw a party. Invitations were sent out, with guests required to wear jackets and ties. The guest list, according to Norman Lincoln, included himself, the Milders, Marvin Taxman, Lawrence Klein, and one or two others.

As everyone gathered at my grandmother's, they were escorted into the dining room, where they were assigned seats at a candlelit table. Norman Lincoln, who grew up in an affluent family, said that he had never seen anything like it: the meal, delivered by servants on English china, began with a series of fanciful hors d'oeuvres, followed by a fish course, squabs, meat course, and ending with fanciful desserts and champagne.

Norman Lincoln remembers my grandmother's reaction to all of this, because she was, for the only time that he could remember, happy.

Norman did not recall my father saying anything about the dinner. If the party had been meaningful to him, he never mentioned it. If he found the party excessive, he never mentioned that either.

A review of my father's high school records bears out Bussy Slosburg's observation that Father had little interest in high school. He played football, belonged to the Motoring Club, and as to his grades, they were average at best. If he was lazy or bored, he was not without ambition. Quite the contrary.

"After Central," Marvin Taxman observed, "Nick wanted to leave Omaha and become a writer. For him, it was clear. He saw himself living in Chicago or New York, writing."

Lawrence Klein, a friend of my father's and one of the

guests at the sixteenth birthday party, remembers my father's desire to leave Omaha and study literature, specifically classical literature. Klein also recalls that this was not what my father's parents had wanted for him; they insisted that he stay at home and follow them into the family optical business. If he wanted to go to college, he could attend Creighton University, down the street from the factory.

In the fall of 1938, Norman Rips left Omaha for the University of Nebraska at Lincoln.

Arriving in Lincoln his freshman year, he moved into a fraternity. Father knew a number of people in the fraternity — friends and brothers of friends — so it made sense that they would recruit him and that he would join.

At the time that he enrolled, there was a strong literary tradition at the school. Willa Cather had studied there. Louise Pound (the sister of Roscoe Pound and the friend and possible lover of Willa Cather) was teaching literature and had a fine reputation as a teacher and scholar.

Here then was the ideal setting for my father: away from his parents and free to study exactly what he wanted. But just months later he returned his suits to the valises and drove back to Omaha, where he enrolled at Creighton University. At the same time, he began to work at the optical factory. Within a year, he dropped out of Creighton. Except for the war, he would never again leave Omaha for any length of time.

This series of events — leaving the University of Nebraska, returning to Omaha, and working in the family business — surprised those who knew him, and various theories have been offered.

That he found the rules of the fraternity confining (there were restrictions on drinking and smoking) is undoubtedly true. He most certainly disliked the intimacy, the forced companionship. But none of this explained why he would have returned to Omaha. There were other colleges, other cities.

It is possible that the answer is found in a remark made by Norman Lincoln. Discussing the wildness of my father and his friends, Norman Lincoln observed that by the time he, Father, and the others were readying themselves for college, there was the expectation that they would all soon be at war. "With the war close," Norman Lincoln told me, "people were willing to let us enjoy ourselves."

Within the community in which my father lived, there was great enthusiasm for the war. Many of his friends were the sons or daughters of immigrants who had been persecuted in Europe, and many of these were Jewish. The politicians and newspapers at the time were strongly behind the war. Jewish leaders were equally enthusiastic.

Following Pearl Harbor, my father's friends in Omaha went to war willingly and with a sense that what they were doing was necessary. In this context it would have been unthinkable for a young man to resist enlistment.

There was one, however, for whom going to war was out of the question, and he seemed, according to his own testimony as well as that of others, to care not a bit about how Jews or patriots or his peers or anyone else for that matter viewed him. That man was my father.

My father's objections to the war were entirely aesthetic: he did not, he told me, like the uniforms, even less the idea of sleeping in a tent; as to eating and sleeping with a group of other men, he found it objectionable; if he had wanted that sort of thing, he would have stayed in the fraternity.

In contrast to this was my father's brother, Leonard, who was four years younger than my father. Leonard and my father shared the same room growing up, went to school together, and from all accounts were close. With the outbreak of war, Leonard fought.

The problem with my father's decision to avoid the war was that there was no easy way out. Healthy and of the right age, my father, everyone was convinced, would end up in the war.

But my father did not go to war.

What people came to learn was that Esther Rips, with the assistance of her lawyers, had convinced the government that the optical factory, of which her son (at a suspiciously young age) was the president, was essential for the war effort. Her argument was no doubt based on the fact that the company, in addition to making lenses for eyeglasses, also manufactured lenses for telescopes, binoculars, and gun sights. If one were looking for an explanation as to why my father abandoned his plans to become a classicist and then a writer, here was one possibility: seeing war ahead of him, he did what he needed to do to escape it.

As his friends left for Europe, my father stayed in Omaha. And this is how it would have remained had the Army not one day received an unusual petition. The petition, signed by a group of women—more specifically, mothers—from Omaha, detailed the exploits of a young man who, while the sons of these women were dying in Europe, was entertaining their daughters with no respect for their chastity. The licentious one was my father.

Nothing the clever Esther Rips or her son could say would change the mind of the Army to restore him to his post at the factory. Father undressed and put on a uniform.

The anticipated time in the tent did not come. To my father's amazement, he was told that he was not going overseas; he was, instead, being sent to college. The Army, in subjecting my father to intelligence tests, had accidentally shaken a cognitive ability that heretofore had not shown itself. Such a man, the Army concluded, would be wasted on the front line; it was in this way that my father ended up in intelligence training.

By 1944, it appeared that the war might end sooner than expected, and my father was pulled out of school and sent to Europe, where he was assigned to a medical corps stationed in Normandy. He treated the wounded and dying and, when called upon, fixed glasses that had been damaged during combat.

On October 8, 1944, his brother Leonard was killed.

My great-aunt Florence put it this way: "Leonard always wanted to visit Europe. The first time he saw it was when he got off the ship; the last time he saw it was when he got off the ship. Leonard floated to Europe."

During the war, Florence would have dinner every Friday night with Esther and Aron. On Friday morning, Florence was told that Leonard was dead. She went over to Esther's that evening. As Florence walked in, she saw Esther "rolling on the floor, screaming." Sheldon said that Esther remained on the floor for three days.

In the numerous letters between my father in Europe and my mother, there are only two references to the death of his brother. One is a matter-of-fact mention of having heard of the death, the second an almost accidental reference to it as a "deep wound." In the years that followed, my father never mentioned his brother Leonard. At the time of my father's death, there was nothing of Leonard's among my father's possessions.

In my father's correspondence with my mother, at the time he heard of his brother's death and chose to say nothing, he does take the time to report on his fascination with Charles Lamb, whose essays Father was reading when not tending to soldiers. Lamb, a friend of Coleridge's, was a highly regarded essayist, whose work appeared in the journals and newspapers of the day. At the age of twenty, Lamb was afflicted with an episode of mental illness. A year later, his sister, suffering from a similar condition, drove a knife into the heart of their mother. To avoid the incarceration of his sister, Lamb assumed responsibility for her, and after a period in an asylum, she was released into his custody.

If my father's letters are devoid of any sign of suffering, so too Lamb's essays, though Lamb does confess that as a child he was "dreadfully alive to nervous terror" and that between the ages of four and seven he never went to sleep without seeing a certain specter. That specter was the Witch of Endor, who in Lamb's vision is pulling Samuel up from the dead.

The sequence then is this: On the front with the wounded and dead, my father, who is told that his brother has just died, takes to reading the essays of Charles Lamb; one of those essays recounts the story of Saul and the Witch of Endor; my father repeats the story to me when I am a boy, around which time he is painting a naked black woman. The story of Saul is lost to me until many decades later when it is mentioned in a coffee shop. Upon hearing the story, I decide to find the black woman.

A block south of where my father had his factory is a used bookstore. In that store I came upon an illustrated edition of the Bible. Turning to the story of Saul, I began to read. The story was familiar, nearly word for word what it was in the edition that my family owned and that was still sitting on the shelf of my mother's apartment. Also familiar was the illustration of the Witch; powerfully built, her face a deep pale, her eyes soft and alarmed—the priestess in the coffee shop.

THE
MAGIC CITY

I.

A TTEMPTING to locate clues that would lead me to the woman in the painting, I had slipped into my father's past. The images my father reading Charles Lamb as his brother was buried, my father staring into the coffin of a slaughtered bootlegger, my father drunk in a bed in the Miller Hotel, made clear to me that the peculiarity of his character had been settled long before he was the "ten-year-old loner," before he moved to Dundee, before he became Nicotine Rips. And yet there was no one alive who knew him from his earliest childhood. I was confronted with the challenge of reconstructing the early life of an unknown man born in an obscure place — South Omaha.

On the cusp of giving up, I received a letter from a person whose existence was entirely unknown to me. His name was Michael Ripps.

Michael had spent a considerable amount of time examining the history of the family and had concluded that the Rips family, which is extremely small, includes Rips and Ripps and that the family comes from the medieval duchy of Baden on the east bank of the Rhine, very near the place where the philosopher Levinas had attended university. From Baden the family had spread out: some to France, others to Prussia, and still others to Holland. There is a town in Holland named Rips, and the names Rips and van Rips appear frequently in Dutch public records dating back to the seventeenth century. In the course of spreading through Europe, members of the family, originally Catholic, adopted other religions; some became Protestants and others Jews (most likely the offspring of a Protestant or Catholic Rips and a Jewish woman).

My grandfather, Aron Rips (from the Jewish Ripses) left Hamburg, Germany, for America in 1907. His first home in America was Clorinda, Iowa, where he worked on a farm. After a short time, he decided to move to Omaha. Since Aron spoke German, he was able to get a job in Omaha with a German oculist. Prior to World War I many of the oculists in Omaha were German or German-speaking. As an apprentice to the oculist, Aron became an expert grinder of glass.

In 1918, Aron met Esther Blumenthal. She was blond, attractive, and born in Omaha. Not long after they met, they decided to marry. Aron was twenty-four.

With Nebraska already dry and Prohibition soon to go into effect nationwide, Aron and Esther married in Kansas City so as to be able to serve liquor at their wedding. Returning to Omaha, life would be easy: Esther's father, Louie Blumenthal, a successful entrepreneur, had promised Aron a half-interest in one of Louie's most profitable businesses—the Miller Hotel.

But the Miller Hotel was not in Aron's future.

Without explanation, Louie retracted his promise.

My grandfather was a gentle person. In the time that I knew him, he spoke harshly of only one person—Louie Blumenthal. At any mention of his father-in-law, the Blumenthal family, or South Omaha, my grandfather would revisit the story of how Louie Blumenthal had cheated him out of the Miller Hotel. The interesting thing about this is that such an about-face was apparently uncharacteristic of Louie Blumenthal—he had a reputation as a forthright man, and as far as I know, no man other than my grandfather had come forth with such a charge. Also, Louie Blumenthal was the first to offer his son-in-law money to set up a new business. For the family, the incident involving the Miller Hotel was transformative—transformative because, as you will see, it sent the family down a course from which it has yet to deviate. Upon my grandfather's passing, the mystery of the Miller Hotel remained unanswered.

2.

IN THE SPRING OF 1881, the Missouri River rose above its banks and swept into Council Bluffs, Iowa, flooding much of the city. When the waters withdrew, the stockyards in Council Bluffs were gone.

Shortly after the flood, men from Omaha surveyed a tract of land located four miles south of Omaha. The land was on top of a rolling hill, well beyond the reach of the Missouri. Pleased with what they saw, the Syndicate, as the group was called, bought 2,000 acres, 200 of which were set aside for stockyards. Stock pens and meatpacking facilities were quickly built, and on May 10, 1885, Omaha society arrived to witness the ceremonial opening of the stockyards. With the crowd gathered, John McShane, the president of the Union Stockyard Company, christened the stockyards by plunging his knife into a hog.

What followed was nearly unimaginable. In its first year in

business, the South Omaha stockyards received 100,000 animals. Eight years later, the number had increased to 2.5 million. By 1900, nearly two million hogs alone passed through South Omaha. The success of the stockyards and slaughterhouses was in part the result of the proximity of South Omaha to the source of supply and in part the function of the rail lines that fed into the city.

On July 18, 1888, a township was registered for the land surrounding the stockyards. The town was to be known as New Edinburgh, in anticipation of money from a businessman in Scotland. The money never arrived and the name was changed to South Omaha.

With the success of the stockyards and slaughterhouses came an expansion of the population of South Omaha. On the land that supported a handful of farmers in 1884, the population rose within two years to 1,500. By 1890, the number of people grew to 8,000; by 1900, it was at 26,000; and it reached 33,000 in 1910. Of these, 31 percent were foreign born.

During this period, South Omaha was the fastest-growing city in America. For this reason, South Omaha became known as the "Magic City."

Since several of the men who formed the Syndicate were themselves immigrants, they had no hesitation in filling the yards with Germans, Russians, Czechs, Italians, Irish, and Danes.

To these were added Japanese. Brought over to break a strike in 1904, they were put up in a mansion, which came to be known as Jap House. Boys from the neighborhood would come to Jap House in the morning and would be treated to breakfast. Japanese continued to live in the house until 1935, when Kiyoshi Ishikawa, who was selling sake and Japanese literature, left the premises.

3.

A FRIEND of mine works at the airport, and it was through him that I met Herb Walker, who headed security at the airport. Since many of those employed at the airport are on work release from the prison down the street, the job is not as easy as it may sound.

In his late seventies, Walker appeared to be no older than fifty; lean and muscular, he had the same head of red hair that he had when he was eighteen. When I met him for lunch, he was wearing a sports coat and tie. He had come from church.

Herb grew up in Omaha. His mother was poor and the only place that she could afford to live was in a part of Omaha that was largely black. As a child, Herb delivered newspapers to the people in the neighborhood.

Herb's mother never married and Herb never knew his father. From what Herb has been able to piece together, his father was Native American.

When America entered World War II, Herb enlisted, though he had to lie about his age to do so. With the war over, Herb Walker went to work in the meatpacking plants in South Omaha. He met a woman from South Omaha; they married and moved into a house in South Omaha, where they had a daughter.

After years in the slaughterhouses, Herb took a job on the Omaha police force. He worked there until he retired. When he and his wife separated, Herb continued to live in South Omaha. To this day, Herb Walker lives in South Omaha, despite the fact that, as he points out, he has saved enough money to buy a place in a nicer part of town.

I asked him why he had stayed in South Omaha, and this is what he told me:

"When my daughter died, I needed a coffin. Not a big one

—two feet—but it was still expensive and I didn't have the money. One day, a man in the neighborhood named Subbie came up to me and asked me what was going on with my daughter. I told Subbie what was going on with my daughter —that she was dead and needed a coffin—and Subbie said to me, 'Herb, you don't worry about this,' and a couple days later my daughter had a coffin and a big tombstone. That's the sort of place you live in when you live in South Omaha."

When Herb moved to South Omaha, he was surrounded by Italian Catholics and started to go to church. One day, in his sixties, he converted to Catholicism. Herb now attends Subbie's church. Subbie, who made tombstones and who brought people from South Omaha to the dead, is now himself deceased.

As a police officer, Herb Walker regularly found himself in South Omaha. "When the trains came in with the cattle," Herb explained, "the women would go down to meet them and then bring them back up to the whorehouses on Twenty-sixth Street."

"Which whorehouses?" I asked him.

"The Plantation House," he responded, "the Miller Hotel . . ."

Herb Walker, without knowing it, had just identified my great-grandmother and great-grandfather as the hostess and host, bawd and whoremaster, of one of South Omaha's preeminent brothels.

South Omaha was particularly open to prostitution. In addition to the large and continuous population of cattlemen who, having spent weeks on the range, sought the entertainment of women when they arrived in town, there were the men in the stockyards and slaughterhouses who were single or had come to America ahead of their families. Other meatpacking cities had larger populations and more diversified industries. South Omaha was nothing but stockyards and slaughterhouses, which meant that South Omaha was nothing but male. In 1886, South Omaha had no jail, no church, and no police force; what South Omaha did have was two whorehouses and a gambling hall.

Once prostitution got under way in South Omaha, it could not be subdued. While there was a certain toleration of prostitution in small cattle towns in the nineteenth century, this ended as the percentage of women increased, as those towns grew less dependent on the cattle business, and as legislators —exercised by the fear of venereal disease and the "white slave trade"—enacted laws to put an end to prostitution. Not so South Omaha, where prostitution flourished into the twentieth century and showed no signs of abating when, in 1915, South Omaha was absorbed into Omaha.

One reason for this was Webber Seavey.

As Omaha's chief of police, Seavey had traveled the country to observe the various means that cities had used to control the sex trade. Seavey concluded that the best way to deal with prostitution was to leave the supervision of prostitutes up to the brothel owners. And this is precisely the system that Seavey put into effect: madams would keep a record of the prostitutes who were in their brothels, check them for diseases, and pay a monthly fee to the city.

Not everyone was pleased with Seavey's views on prostitution and on more than one occasion he was called before city officials to explain himself. Responding to the suggestion that prostitutes should be severely punished, Seavey could not have been more contemptuous:

"Degraded in their own estimation, and heartbroken, the

fallen women find no pity, mercy, or sympathy among their more fortunate sisters who have never made any known mistakes, and they shrink with fear and shame from parental meeting, and as the last resort naturally enter the doors that alone are open to receive them and enter the ranks of sisterhood that straight-laced virtue unpityingly and falsely designates as a lost woman."

As to who the real malefactors were, Seavey had no doubt.

"Men who destroy confiding girls with as little pity or mercy as the tiger destroys the lamb are received with open arms by respectable girls, who regard the crimes of these human vampires as little masculine mistakes to be overlooked, of course, in those young gentlemen, who, instead of remorse for their horrible acts, feel proud of their loathsome conquests and glory in their shame."

Public records and newspaper articles reveal that though there was prostitution throughout South Omaha, the heart of it, as Herb Walker knew, was located around Twenty-sixth and N streets. George Sedlacek, a reporter who covered South Omaha in the 1920s, observed that "every building between Twenty-fifth and Twenty-sixth Street that could be put to that kind of use was a bawdy house at one time or another." This included, in addition to the Miller, the Plantation House (also known as the House of All Nations, for its claim to offer women from all nationalities and races), the Atlantic Hotel, and Gus's Mad House. Among the pimps and madams was Willie Counts, who ran a speakeasy for blacks, Lightning Johnson, Porcupine Jim, Madam Smeal, and my great-grandmother and great-grandfather—Anne and Louie Blumenthal.

Prohibition had given rise to an increase in prostitution, and it was at precisely this time that my great-grandparents decided to get into the business by taking over the Miller Hotel, which they owned until 1942. By that time, Louie Blumenthal was dead and his wife had enough money not to have to worry about running a whorehouse.

The Miller Hotel continued as it was, and as late as October 1972 the Omaha *World Herald* reported the arrest of one Tessa Gallagher for prostitution after several men entered her room at the hotel. Two months later, the city of Omaha sought an order from a judge to padlock the Miller as a public nuisance, with the city prosecutor charging that "the premises are maintained for the purpose of allowing hoodlums, prostitutes, gamblers, and other disorderly persons to gather."

Jack Kawa, the owner of Johnny's Café, a steak house around the corner from the Miller, tells the story of how he and his friends would go up to the bar at the Miller to take in the crowd — cowboys, prostitutes, businessmen, and workers from the slaughterhouses. But of all these his favorite was Tanya, a habitué of the bar at the Miller. What distinguished Tanya from others in the bars of South Omaha was that Tanya wore on one half of his body a tuxedo, cropped hair, and an unpainted face, and on the other side of her body a dress and heels, long hair, and makeup. Depending on whom he or she was seducing, Tanya would sit to one side or the other, sometimes both.

Jack describes one evening when Lightning Johnson — who ran the prostitutes on Twenty-sixth Street and drove around South Omaha with a bag of money and a Luger on the front seat of his car — decided that his business, having fallen off, needed some promotion. Johnson gathered the prostitutes from the Miller and other establishments and marched them through the center of South Omaha. At one point Jack looked out the window of Johnny's and saw the prostitutes, dressed up and waving their arms, circling his restaurant. At the front of the parade was Lightning.

The next day, Jack found Lightning. "Lightning," he said, "your domain is over there—the Miller Hotel; mine is Johnny's. If I see another parade, I'm going to get out my shotgun and start shooting hookers. You're a friend, so I'm not going to shoot you. I expect the same courtesy." There were no more parades. A year or two later, Lightning was imprisoned overnight for a traffic violation. It was the first time that he had been arrested. That night, in the jail cell, he passed away.

It was in this atmosphere that my great-grandmother and great-grandfather ran their brothel. And if this atmosphere was special it was made especially so by its fragrance. Backed up against the stockyards and slaughterhouses, the Miller Hotel was encased in the aroma of blood, feces, and animal feed. There was no escaping it, and no reason to escape it. Father had told me that my great-grandmother loved that odor, and at some point in every day she would leave the Miller Hotel to freshen herself with a noseful of it. Another resident of South Omaha said that when he smelled feces he smelled money.

At Midway Plaisance in South Omaha, an organ concert was hosted by Signora Rita Piselli, and if my great-grandmother didn't attend the concert, it was certainly something that she would have enjoyed. The first notes that greeted the audience from the organ were burning tar—the tar of the mastheads of ships at bay in Naples, announced Signora Piselli. The organist played on—the pipes of his organ stuffed with perfumes, incenses, and attars. Now there was the scent of the carriage as the lovers drove from the bay to the church. And so it continued as Signora Piselli took her audience through the scenes and scents of the wedding of the young Italian lovers—the church (lilacs and talc), the honeymoon (one could only imagine), breakfast the day after (bacon and eggs), and finally a stroll in the country (new mown hay).

When Louie Blumenthal told my grandfather that he would not be a partner in the Miller Hotel, my grandfather retreated to the one thing that he knew, grinding lenses, which he had learned from the German oculist.

Aron and his wife, Esther, set up a business in which they filled orders for local optometrists, with Aron working in the factory and Esther doing the bookkeeping. At the end of the day, she collected the finished orders and delivered them to different parts of South Omaha and Omaha. For the orders that went to Omaha, grandmother rode the trolley that ran from Twenty-fourth and N streets to the corner of Twenty-second Street and Ames Avenue.

The mystery surrounding Louie Blumenthal—his decision to retract his promise on the Miller Hotel—may well be explained by Louie's having second thoughts about getting his son-in-law, a religious Jew, involved in a brothel.

The birth of my father in 1920 presented a difficulty for his parents. Since both of them worked long hours, there was no one at home to care for their son. The answer was to leave my father with his grandparents, who, if they could not take care of the child themselves, had a house full of young women, who were happy to look after him. And this is how it came to pass that my father was raised in a brothel.

The definitive word on this came from Florence Blumenthal, who was married to Louie's son, Russell. Florence is the only one left in our family of her generation, so if she was ever inclined to hold her tongue, she is no longer. When I mentioned to her the possibility that there were whores at the Miller, she responded, "Of course there were whores at the Miller, it was a whorehouse. Anne told me that she left the business when Louie got sick, but the truth is that she would still be there with the hookers if business hadn't fallen off during the war."

4.

THE MILLER is now gone. So too the other brothels of South Omaha.

The buildings that housed the nineteenth- and early-twentieth-century slaughterhouses are, with one exception, also extinct. That exception is the Wilson plant, a red brick, 500,000-square-foot, six-story building, with a 190-foot smokestack, which sits on a hill overlooking the empty stockyards.

Since 1976, the Wilson slaughterhouse has been closed up: walls collapsed, windows shattered, the building filled with rats and other animals, which, along with water and snow, occasionally awaken the blood on the walls. The city condemned the building, but before it was to be destroyed something happened: an elderly man appeared from inside the building. His name was Otis Glebe.

Otis had been living in the building, without heat, water, or light, for many years.

When the city let him know that they intended to demolish the building, Otis Glebe went to court and argued that a man is entitled to his home and it mattered not that his home was full of carcasses. The court agreed and Otis Glebe was pronounced the rightful occupant of the slaughterhouse.

At some point in the proceedings between Otis Glebe and the city, Otis Glebe was forced to live in the dirt next to the building. If he needed water, he would have to leave the property and find a public water fountain. If he needed a bathroom, I'm not sure what he did. According to Otis, he dined on "whatever he could find." As to that, he was not specific.

Determined to destroy the plant and frustrated by the court's ruling that the slaughterhouse was Otis's home, the city moved to take his house in an eminent domain proceeding. The city's efforts were successful and Otis was locked out of the slaughterhouse.

Otis Glebe's efforts to preserve the slaughterhouse gathered few supporters. The slaughtering of animals was now less important to the economy of Omaha, and Otis himself was a disturbing figure.

Otis owned buildings that he rented out to tenants, who shortly after moving in discovered that they had no water or heat. Otis explained that he did not believe that it was his obligation to provide heat or water to his tenants—he was, after all, living in the dirt outside a slaughterhouse. His tenants complained to the city, and the city dispossessed Otis of his buildings.

Standing before the locked gate of the slaughterhouse, Otis Glebe told me that he wanted nothing more than to be back inside the slaughterhouse; he missed his live animals, dead animals, and memories.

Otis also told me that his interest in the Wilson plant began when he worked at a building across from the packing plant. He could not stop watching it: in the morning as the cattle entered the building, it began to bulge from the thickness of the slaughter. This lasted until the evening, when the building could consume no more. As Otis grew older, the slaughterhouse returned to him: he dreamt about it, heard it beating, began to visit it, and, when it closed, thought of how he might make it his own.

The ancients associated blood with life and consequently believed that the blood of a slaughtered beast would give strength to a human or even revive the dead (Odysseus filled a trench with blood to allow the shades to drink). "In the ritual of shedding of blood," E. O. James wrote, "it is not the taking of life but the giving of life that really is fundamental, for blood is not death but life." In Sophocles' *Ajax,* animals are substituted to save the life of Odysseus, and in the first book of Samuel, David's wife used the hair of a goat to give David's assassins the impression that he was still in bed, thus saving his life.

The day was growing late, and I'd not eaten. I knew of a place near the slaughterhouse. I asked Otis to join me. He declined. He was an old man, he told me; he would remain with his building.

5.

THE RESTAURANT was in an area full of small houses—the houses that were once filled with workers from the slaughterhouses. Herb Walker lives in one of these houses; my father was born in another; today a greater share of the houses are occupied by Hispanics. Some work in what is left of the meatpacking business. Others work in the body shops, taquerias, and schools.

The house that I entered, no bigger than the others and covered in white stucco, is owned by Yvonne Kuntzel. The road in front of the house is lined with pickup trucks. Across from the house is the River City Barricade Co. and down the street Tony and Judy's Baratta's Again Bar.

Thirty-four years ago, Kuntzel, who was born in Serbia and escaped from a refugee camp in Trieste, was working in Washington, D.C., when he witnessed something that changed his life. Watching television on a weekend afternoon, he saw animals in their natural habitat.

"I couldn't get it out of my mind—wrestling animals, pushing animals. I said to myself, who put this on television— wrestling animals, pushing animals—and when I saw it was something called Mutual of Omaha, I thought that is where I should go, Omaha."

"People said to me 'Yvonne, you are crazy! You can't go to Omaha.' So I packed up and went to Omaha."

The front door of Yvonne's house opens onto the living room, but there are no couches or bureaus, no bookcase or piano. Only tables. For it is here that Yvonne serves his meals and that a knowing few come to eat.

At Yvonne's it is pointless to appear at the door without reservations; Yvonne admits no one who has not had the good manners to call ahead of time. Yvonne also insists that gentlemen wear coats and that women dress appropriately. But these only foreshadow the constraints that bind one to the meal, for Yvonne directs every decision that the diner makes, much as an expert equestrian guides his horse about the rink with no apparent movement of his arms or legs. In this way Yvonne is Antoine Beauvilliers, the proprietor of the eighteenth-century Grande Taverne de Londres, of whom Brillat-Savarin wrote that he "point[s] out here a dish to be avoided, there one to be ordered instantly . . . and send[s], at the same time, for wine from the cellar, the key of which he produced from his own pocket; in a word, he assumed so gracious and engaging a tone that all these extra articles seemed so many favors conferred by him."

As one waits for the meal to arrive, the busboy, who has been with Yvonne for as long as either of them can remember and is now in his sixties, stands at attention next to the table. In the opposite corner of the room, Yvonne paces back and forth, his head bent, his face suffused with thought. He is concerned that your meal is as you want it, but he's more concerned that it is as he wants it. And it is this ideal, I suspect, that is his agony.

The arrival of the meal is too much for Yvonne to bear.

The meal is exceptional.

He bounds to the table, his arms wagging.

"You like, Madame?"

He knows the answer.

Watching him, I notice a portrait of the Queen of Sweden, which I point out to my wife. The queen wears a full-length blue gown and a gold crown. Yvonne notices my gaze.

"Monsieur, you like?"

"Yes, I do; it's a great likeness."

"The Queen of Aksarben, a very good customer."

Aksarben is Nebraska spelled backwards and refers to a charitable organization that holds a ball every year to raise money. Aksarben used to sponsor a rodeo (in which I participated as a child) but they have abandoned that. Today Aksarben elects a king and queen, who appear in crowns and robes and march around an arena. The queen is chosen from among the most upstanding young ladies of Nebraska.

There are other customers on the walls, including Ike Friedman, a beloved jeweler. There is also a sketch of a man who went by the initial "X." and was a cousin of my father's.

X. first came to my attention upon the engagement of his daughter. The girl had fallen in love with another member of our family, and they had decided to marry. At the time of the engagement, the daughter was not old enough to marry and needed the consent of her parents. X. said no.

X.'s wife intervened: she told her daughter that if she was patient, X. could be prevailed upon to change his mind. After a few months, X.'s wife became convinced that X. was ready to give his consent, and invitations were sent out for the wedding. The reception was to be held at Yvonne's restaurant—the Café de Paris.

Shortly before the dinner, the wedding was canceled. Contrary to what his wife had thought, X. would not agree to the marriage. The daughter and her boyfriend ran away. En route to wherever they were going, they were in an accident and the young man suffered a blow to the head that left him in a coma.

X.'s wife wanted no more of X.

She moved out and set up house for herself and her daughter. This was followed by a ruling from the judge on the terms of the divorce and a statement from X. to the effect that he would not pay the amount that the judge had ordered. X. would not pay any amount.

This was not to the judge's liking and he sent X. to jail.

From his jail cell, X. issued manifestoes on behalf of all men who were unfairly treated by their wives. X. gained publicity from this and there were even some who gathered at the jail in his support. None of this impressed the judge, who was perfectly content to leave X. in jail. And this is when X. hired one of America's most famous trial attorneys to come and get him out. The attorney had never appeared in Omaha and his arrival was carried in the local papers.

The famous lawyer assured the court that the whole matter was a great mistake and that X. was sorry for any confusion he had caused; X. would be happy to pay the amount mandated by the judge. X. was released but when the time came to pay, he had vanished.

The refusal to pay the alimony, the speeches about the unfair treatment of men, the hiring of the famous attorney, the time that X. had spent in prison were all part of a charade: X. was buying time while his elderly mother secreted X.'s assets out of the country. She apparently carried them in her panties, for a friend of mine who ran into her at the airport reported that X.'s mother had a bulge near her groin that shifted from side to side as she walked.

There are rumors that X. slips in and out of the United States. A fellow from Omaha claims that he saw X. on a golf course in Florida but that when the fellow ran to phone the police, X. spotted him, surmised what he was doing, and drove his golf cart off the course, onto the highway, and from there to the airport.

To my father's mind, X. was a buffoon, though this understates it. To my father's mind, X. had been so shrunken by avarice that he was no longer visible. As my sister-in-law Jane has pointed out, my father was an ethical man: he took care of his family, was kind to those who worked for him, helped those who were in need, and made his dislike known (although never verbally) of those who, like X., fell short of this.

During his life, my father would have nothing to do with X. Yet sitting at Yvonne's I could not help but notice the proximity of the bad X. to the good queen; and of these two to Otis, his slaughterhouse, and the whorehouse of my great-grandparents; and of these three to my father, who was born in the Magic City, to Herb Walker, who returned there, and to his daughter, who never left, and to Subbie, who took people to the underworld.

Months after returning to New York, I was back in the coffee shop. The Bearded Priest had reappeared. On his table was a copy of Levinas's *Totality and Infinity*.

Following in the path of Husserl and Heidegger, Levinas was committed to uncovering the constituents of consciousness. Believing, as did Husserl and Heidegger, that consciousness was thoroughly embedded in the world, he stripped aside all presumptions, including those concerning the real as well as metaphysical worlds, including the presumption that there was a division between the subjective mind of man and the objective world beyond that mind, to discover—the discovery set forth in *Totality and Infinity*—that the essence of man's mind was his awareness of the Other.

The Other, according to Levinas, cannot be known by man —a transcendence that provides no guidance or salvation. Levinas uses the term "Other" interchangeably with "Infinity" or "God" or the "Mystery of God," understanding that no term is adequate given the incomprehensibility of the subject.

Critical to Levinas's philosophy is the idea that man's awareness that he is radically apart from the Other (man is not an extension of the Other, nor is the Other an extension of man) confers existence on man by way of contrast and hence places man in the debt of the Other.

In the face of other people, man becomes aware of the Other and, consequently, the debt owed to the Other. The term "face," as used by Levinas, means both the material face of other people but also the unapproachable Infinity or Mystery that is reflected in the face.

"The face," Levinas observes, "expresses itself in the sensible, . . . [but] the face tears apart the sensible." Here Levinas repeats a story told by Vassily Grossman of how in Moscow, before a gate where people were allowed to drop off letters for friends and relatives arrested for political crimes, "people formed a line, each reading on the nape of the person in front of him the feelings and hopes of his misery."

Engagement with an Other that is never completely revealed, that is always the subject of contemplation, changes us, Levinas believed—it allows us to develop, freeing us from the illusion of a "true self." To reflect on it was to "explode" consciousness, not save it. In this, Levinas distinguished himself from Heidegger, who held that the forces exterior to man's mind were entrapping and that man could only find freedom by pushing aside those forces, and in doing so, allow an authentic self to emerge.

I left the Bearded Priest and returned home. Waiting for me was a letter from Omaha. Inside was the obituary of Otis Glebe.

Otis was not in the building when it was blown apart. But the effect was the same: shortly after its demolition, Otis died. His lawyer was quoted in the papers as saying that the destruction of the slaughterhouse had killed Otis.

As I related the story to the Bearded Priest, he nodded. A glebe is the land owned by the church and given to a priest. The priest grew crops on the glebe and in this way supported himself. The slaughterhouse was Otis's glebe.

Over a year later, it was announced that Otis Glebe had died with an estate worth $4.5 million. Of that amount, he left $3.5 million to charity, including enough money to a soup kitchen in South Omaha to provide food for a year.

Upon reading Levinas's translation of Husserl, Sartre's vision of the world changed—he became aware, he wrote, that "truth drags through the street, in the factories, and, apart from ancient Greece, philosophers are eunuchs who never open their doors to it."

THE CIRCUS

ANNE MILDER, the only sister and youngest sibling of Morris Milder, was lively and pretty and happy.

After graduating from Central High School, she met Benjamin Taxman. Ben's father, J.J., knew the Milders, and when he was passing through Omaha he would keep his eye out for young women who might be suitable for his sons. He thought Anne would be right for his son Phillip, but it was Ben who fell in love with her.

Ben was the perfect consort. Himself from a wealthy family (the Taxmans had made their money in the oil business), Ben was polite, soft-spoken, and, like his father and brothers, favored fine clothing, large meals, and the other pleasures that the gentlemen of their day enjoyed. Ben and Anne would marry, have a daughter, and one day much later that daughter would be my mother.

Ben and his cousin, Doc Seigal, used to meet regularly for lunch in the lobby of a restaurant in Chicago. In the middle of one of their meals, Doc excused himself from the table. Ben asked where Doc was going, and Doc said he had business to take care of but would be back shortly. Ben was suspicious: Doc had never before left the table for business, and besides, Doc had no business. Ben decided to follow him.

Ben watched as Doc headed back into the hotel, climbed the steps to the second floor, and proceeded down the hallway. Doc stopped at a service closet and disappeared inside. When he came out, he was not himself—he was a much greater version of himself, having squeezed two pillows into his suit pants. His jacket strained to contain the bulge.

Minutes later, he entered the ballroom on the second floor. In the ballroom was a conference of bankers. He knew no one at the meeting, though he made a point of introducing himself to as many people as possible. After a half-hour, he left the room.

Ben hurried down the hall and back into the restaurant, making sure that Doc did not see him. Minutes later, Doc arrived—without the pillows. Though curious, Ben said nothing. In a few months, the incident was forgotten.

It was almost a year later that the two of them were having lunch in the same hotel, and Ben noticed a sign announcing the conference of bankers that had been there the day Doc had stuffed the pillows into his suit. Saying nothing, Ben continued with his lunch, but before they had finished, Doc excused himself from the table. Again, Ben followed him. Doc made his way to the second floor and again walked down the hallway toward the ballroom, but this time he did not stop at the service closet.

As Doc walked into the ballroom, everyone went silent. Then they began to approach him. So thick was the crowd around Doc that he was no longer visible to Ben.

Ben drew to the edge of the crowd.

At first, Doc demurred.

The men pleaded.

Doc would not speak.

They continued to press him.

At last, he gave in. He would talk, but only if they agreed to keep what he said within the room.

What followed was the story of a miracle. To be precise, a miracle invention. A device that allowed a man to eat exactly what he wanted and still lose weight.

What was the invention?

"Reducing underwear," Doc whispered.

Where could they find them?

"I know of only one pair," Doc explained. "The inventor died before bringing the rest to market." Tears neared his eyes.

Where could they find that pair?

Doc glanced downward.

What began then was an impromptu, perhaps unprecedented, auction for the underpants of the auctioneer.

Following their marriage, my grandparents, Ben and Anne, lived in Eldorado, Kansas, where his family owned an oil refinery. There they had a son: the Marvin Taxman who would one day read Haliburton with my father. By the time my mother (Barbara Joy) was born, Ben and Anne had moved to Kansas City. The two children were placed in private schools and raised by a governess named Elvira Meskimens. Born in Onega, Kansas, she was of English descent.

With the stock market crash of 1929, Ben lost his fortune. He was not alone. His brothers had also suffered.

Two in the family survived: Ben's father, who had invested conservatively, and Ben's youngest brother, whose money had gone into a trust account owing to his age. Ben's youngest brother was driven from Kansas City to Harvard in a chauffeured limousine. With neither the taste nor economic incentive for hard work, he was driven back.

By the time my mother entered high school, Ben and Anne had moved to Omaha. There, my mother and her brother were enrolled at Central. There, my mother and her brother met Nick Rips.

Barbara Taxman had a disciplined intelligence, was well mannered, and had an interest in literature and the arts. This appealed to my father and he began to take her out. At the age of sixteen, she entered Northwestern University, where she studied chemistry.

My father visited her at school, and after the war they were married.

Following the wedding, Nick and Barbara lived among the very people with whom my father had gone to high school. There were no blacks in the neighborhood, no poor, no one but those like themselves.

During the day, the men worked. Many, such as my father, had taken over family businesses; others had obtained professional degrees and set up their practices. When my father was not working, he was out with my mother or with my mother and other couples. Every night they would meet at the restaurants or clubs of Omaha and there would dine and drink, smoke and act in ways that were sexually suggestive.

Many of the steak houses which they frequented have closed. Of those still open, Gorats is arguably the most well known. My father was a regular at Gorats.

The front door of Gorats opens onto a long hallway, which leads to the hostess. The hallway is lined with photographs of cows. Every year prizes are given to the young man or woman who has raised the finest cow. The youngsters in the 4-H Club spend a great deal of thought, effort, and affection in raising their beasts.

Following the prizes, the winning animals are auctioned. Of the bidders, the most interested are the steak houses. The reason for this is that it is prestigious for a steak house to have purchased the year's top cow. The photographs on the walls of Gorats are photographs of the owners of Gorats and an ecstatic youngster standing beside the steer that the owners of Gorats had won at auction. Ross's Steak House went one step further, actually placing the live cow outside the restaurant, where it would wander around in a pen before it was taken inside to be sliced up.

Johnny's Café had neither the photographs nor the perp walk but did feature testicles ("rooster fries") as a bar snack. Johnny's, which is where Lightning Johnson marched his prostitutes, opened early in the morning, serving breakfast to those who worked in the slaughterhouses. The workers from the yards and slaughterhouses would enter through the basement, where they would hose off before going upstairs to the restaurant.

Weekends were spent at the Highland Country Club, where my father would play golf. In between the front and back nine, he would have lunch at the clubhouse. After golf, he might have a drink and a smoke before returning home. Below the bar was the locker room, where my father kept his shoes and jacket. My father's locker had a combination, and there, I imagined, would be a good place to keep a gun. Instead, he kept his gun at the office.

The wives of these men, including my mother, did not work. They woke up, organized their new homes, and, having finished with that, would meet their friends, with whom they had dined and drunk the night before; and they would shop.

The place to shop was the Brandeis Department Store. The French Room, as managed by Madame Flo La Boschin, was where the ladies of Omaha bought their finest dresses. Done buying dresses, they would take their lunch in the dining room on the tenth floor. From the restaurant, they could watch as a woman, grasping a rope with her teeth and only her teeth, floated from the roof of the department store to the sidewalk across the street.

The couples with whom my parents associated began to have children. My parents had their first son, Lance, at the end of 1948.

2.

M Y FATHER had returned from Europe two years earlier. By that time, my grandmother's screams, the screams for her son Leonard, the son who had died at war, had passed to the mouth of another mother, and now my grandmother was quiet. It was not long before my grandparents announced that they no longer wanted to live in Omaha. They wanted to leave and they did: Esther and Aron boxed up what they needed, and with their son Sheldon and Esther's sister in the backseat, they headed south.

There are various theories regarding the cause of their departure. Here is one: my grandparents, upon the death of Leonard, did not want to see him in the sympathetic faces of their neighbors, and in Omaha everyone was their neighbor.

Here is another: neither Aron nor Esther was so attached to Omaha that they would stay if presented with the opportunity to go, and that opportunity came when their son Norman returned from the war. He would run the factory; they would retire.

And a final explanation: Norman Rips returned to Omaha with the idea of running the business and did not want his parents, particularly his mother, around. It was he, according to one observer, who pushed them out of Omaha. There are two flaws in this: it ascribes a sinister side to my father that so far as I could see appeared nowhere else in his life, and it assumes that he, or anyone else, was capable of forcing Esther Rips to do something she did not want to do. It is also belied by subsequent events.

In Florida, Aron and Esther found a town that they liked, settled into a house, and enrolled Sheldon in elementary school. But as soon as all the preparations for their new life in Florida were complete, they returned to Omaha. The house in Florida was sold, Sheldon was withdrawn from school, and they all drove back by the same route they had taken down.

Aron and Esther did not discuss the reason for their return, and today it remains unknown. What can be said is that the decision was most certainly Esther's; Aron, a passive man, made no decisions. My mother's speculation is that Sheldon was miserable in Florida, and Esther decided that it was not right to keep him there.

It may also have been that Esther was miserable in Florida and decided that it was not right to keep herself there; having withdrawn from everything that reminded her of her son Leonard, she found that she missed Omaha. This theory finds support in the fact that upon returning to Omaha, she surrounded herself with family: buying neighboring houses on Thirty-eighth Avenue, she moved herself and Aron into one of them, convinced my father and mother to move into another, and settled her mother around the corner.

For my parents, the new house offered the attraction of being larger than the apartment in which they had been living. It was also in a nicer neighborhood. Filled with mansions and near the Blackstone Hotel, the neighborhood was referred to as the Gold Coast.

Shortly after my parents moved into the house on Thirty-eighth Avenue, their second son, my brother Harlan, was born. If my father had had any enthusiasm for a family, it had now exhausted itself. My mother, though, had different plans. She wanted a girl.

So certain was she of this, so certain that she would have one, that she had early on begun to purchase clothes, toys, blankets, little bracelets, and other feminine finery for that daughter, all of which was shelved in a separate closet. Each pregnancy gave her new hope and new opportunity to add to the inventory.

As planned, three years after Harlan was born, six years after Lance was born, my mother was pregnant with the daughter that would complete the family.

On a weekend in May, with my father playing golf, mother went into labor. The delivery took place at St. Joseph's Hospital. Father arrived in time for the birth of the child.

When my mother was told that it was another boy, she wept.

3.

WITH MY ARRIVAL, Norman and Barbara decided they needed a new home. That new home was built on Sixty-ninth Street, near the edge of the city. A dirt road ran in front of the house, and across the street was a field.

Not far from the house was a stable, and it was there that I learned to ride, under the tutelage of a woman named Mrs. Wolf.

Every year there was a rodeo in Omaha, and Mrs. Wolf

trained her students, including myself, for that event. Cowboys came from all over the Midwest to participate in the rodeo.

I had a chance to meet some of the cowboys, who would linger outside the auditorium while waiting for their events. On one occasion I was standing next to the horse that I was to ride that day and one of the cowboys came over to chat.

"Let's have a look at that horse," he said.

I was young and was happy to have the attention. The horse was one that Mrs. Wolf had picked out for me, and I had grown fond of him. He was white with brown spots and had a warm manner. His name was Little Billy.

The cowboy approached the horse, looked into its eyes, moved its head back and forth, checked its feet, and finally went around to have a look at its hindquarters. He knew that whatever this horse and I were to do in the rodeo was inconsequential, but he feigned seriousness. He and I were in this together, fellow cowboys.

When he got to the back of the horse, the pretense fell from his face.

He was staring at the unusually large scrotum of Little Billy.

"You're a lucky cowboy," he said.

Then, carefully and with evident experience, he removed his glove and, extending his forearm, allowed his fingers to slide back and forth across Little Billy's ample sac, giving rise to a look of pleasured confusion on the face of Little Billy.

It was on Sixty-ninth Street that Barbara Rips had her fourth and final child—her fourth and final son, Bruce. With Bruce, it became clear to my mother that she was never going to have a daughter, and she stopped trying.

Elvira Meskimens, having raised my mother, was now employed by my mother as a governess. She knew my mother well and, sensing her despondence, hit upon a solution: if my mother could not give birth to a daughter, Skimmy (our name for Elvira Meskimens) would find one.

The little girl she found was named Gretchen.

Gretchen had dark skin, brown eyes, and wavy hair. She was attached to mother but was particularly close to Skimmy, whom Gretchen followed around night and day. Skimmy dressed her in the morning from the girl's clothes that Mother had been storing and made certain that she always had a ribbon in her hair.

The effect of Gretchen on my mother was completely curative: Gretchen satisfied my mother's craving for a little girl and even led my mother to the conclusion that little girls were not all that different from little boys. And when a few years later Gretchen left us, my mother showed no regret.

My own memories of Gretchen are faint.

About this same time, my mother's mother (Anne Taxman, née Milder) began spending time around our house. Her husband, Ben Taxman, stayed in Centralia, Illinois, where he and his brothers ran a refinery.

Refineries, according to my brother Harlan, are the ideal business for the indolent man, since they pretty much run themselves. My grandfather and his brothers spent a lot of time reading the *Racing Form* and taking naps. Their office had a couple of desks and a couch. When I asked my grandfather what he and his brothers did at the office, he replied, "In the morning we debate over who will get the couch in the afternoon."

Relatives would make the trip to Centralia to visit my grandfather and great-uncles. Despite the fact that their lives had fallen quite a ways from where they had once been, they remained wry and entertaining. While my brothers and I would stay with my grandparents, others, on the recommendation of my grandfather and his brothers, would be booked into the Pittenger Hotel. The Pittenger, owned by Bertha and Leo Levy, was a short drive from the refinery.

Buddy Taxman, who was staying in the Pittenger while visiting his relatives, was dressing for dinner when he discovered that his ties were missing. What was strange was that several years before, when staying at the Pittenger, the same thing had happened.

He reported the theft to the Levys, who told him that they would look into the matter. Later that day, he mentioned what had happened to my grandfather and his brothers. They explained that there was no need for concern. The ties had been stolen by Bertha, who was a kleptomaniac. Buddy picked up the phone and called the hotel, demanding that the ties be returned to his room. Bertha, who could not have been more considerate, agreed.

When Buddy finally got back to the hotel, he found that Bertha had, as she had promised, returned the ties, with each folded neatly on the bed—the ties from three years before.

These amusements were evidently not enough to satisfy my grandmother, and she began to spend increasing amounts of time in Omaha. As I was growing up, she was living in our house a greater part of the year. Occasionally she would be joined by my grandfather.

When grandfather came from Centralia to visit his wife, he would wake up late, eat a lunch of liverwurst and cheese, and then take his constitutional. If I happened to be around, he would inquire whether I wanted to go outside and catch some snakes.

Armed with a dandelion picker in one hand and a cigar in the other, he would lead me around looking for snakes. His eyes were weak, but for snakes he could see through trees, deep into bushes, down small holes. I was never fast enough to catch a snake or even see one, but that did not stop me from trying. Grandfather would suddenly stop and point to the snake, and I would charge across the fields with the dandelion picker on my shoulder. As I turned back, not having found the snake, he hid his flask. If it began to snow, he would say, "Michael, it's colder than a mother-in-law's kiss," and lead me back to the house.

On those occasions when my grandparents stayed with us in Omaha, they would often be visited by the brothers of my grandfather. My great-uncles would drink and tell stories about life in the small towns of the Midwest and of their adventures as young men.

With four sons, governess, husband, mother, and sometimes father and uncles in the house, my mother had less time than she needed. On top of this, she felt the obligation of various causes, to which she devoted considerable effort.

The demands on my mother meant that I spent a greater share of my time with Skimmy. Skimmy was in her seventies and wore ankle-length white dresses, thick stockings, and starched aprons. Her hair was a luminous white and was pulled back into a bun. Her face was at once soft and serious. Her horn-rimmed glasses suggested a poet.

I followed her around the house. When she sat down, I was in her arms; when she went to her room, I was behind her. For me, she was tranquillity.

Each of us was assigned to Skimmy's room when we were first brought home. She gave us our bottles in the evening and took care of other needs. With the arrival of a new child, we were moved into our own rooms.

For some reason this did not happen to me. The room my parents wanted to move me into was full of bright colors, but I did not want to be moved and made such a fuss about it that I was left with Skimmy. If she minded, she was kind enough not to say anything.

From my bed, which was next to hers, I watched each morning as she prepared for the day. At night she would sit erect in the chair across the room and read a book. She was not a religious woman, but her face had the quiet of those whom I had seen in prayer.

One spring day, I was delighted to find that Skimmy was still in bed. She had decided to take a few days off, which she did from time to time.

I climbed out of bed to wake her but decided against it. Instead I touched her hair, in the manner in which she touched mine. I climbed into bed next to her, trying not to disturb her, touching her again.

What my parents saw when they came into the room was not what they thought. They thought they saw me in bed with a corpse. Though this was true, it did not allow for the possibility that Skimmy had permitted me to be with her in her new home, Death, undisturbed by bereavement.

That morning my mother made breakfast and sent me off to school. Arriving back that afternoon, I returned to Skimmy's room but the door was closed. I opened the door, slipped into the room, and lay back down on the sheets. She did not come. So that she could find me, I moved the creases and shadows into the position that they had been in when she died.

4.

WITH HIS SONS growing up, Father assumed an amiable apathy that he would maintain without interruption for the rest of his life. At no point did he think to play with his sons, take them out in the yard, teach them a sport, attend school activities, or, as we grew older, inquire about our jobs or even the women whom we dated or married. He was there with us but was transparent. A sort of spirit—at least that category of spirit that wears bespoke suits and tells stories about antiquity.

If my mother, grandparents, or anyone else was troubled by my father's abdication of his duties, there was no mention of it. Lance and Harlan helped me and there were fathers in the neighborhood who were happy to do what they could. Bob Wagner, who was full of knowledge about electronics, was willing to give me guidance on that subject. Gary Gross, our next-door neighbor, was an insurance man, but he had been a trumpet prodigy and seemed to know a lot about the world on top of that.

There was only one time that I remember Mother insisting that my father fulfill his obligations. The circus had come to Omaha and Mother thought that I would enjoy it. Because the circus was sponsored by the Shriners, several of whom were acquaintances of my father, they would certainly be in attendance.

I am not certain what Mother did to convince Father to take me to the circus, but over dinner she announced that I would be going with him and that I would be allowed to invite a friend or two. This was my first circus and I did not know what to expect, my only knowledge coming from a book—*Doctor Dan at the Circus*—that my friends and I had passed around. In that story, a group of kids got together to create a circus. The kids assumed different parts. One of the children, Dan, insisted that he wanted to be the circus doctor. The other children scoffed. But Dan was right to want to be a doctor: during the circus, a couple of the kids fell down, and he was able to fix them up.

Given my limited knowledge of the circus and with the big

day approaching, I walked over to chat with Gary. His son, David, had told me that Gary knew about the Shriners.

Gary was sitting on the couch of his living room, smoking cigarettes, a soigné expression on his face.

I told him that I was going to the circus.

He released a doughnut of smoke from his lungs, waited till it crossed the room, and then began to speak, giving the impression that his thoughts were attached to the doughnut, which when exhaled pulled them to consciousness.

What Gary told me that afternoon was that the Shriners were a society of ancient Arabs that had come into existence after a night in France when people got drunk, smoked opium, and came under the spell of a man who was never visible but wore a fez and brassiere, which is how people knew he was present. Gary went on to explain that I should not be frightened if the men at the circus were wearing purple hats and matching brassieres.

Gary had made sense of the circus, so when the day arrived I felt prepared. I gathered two friends, did a little reading about circuses, and packed my copy of *Doctor Dan at the Circus*. I was excited about the circus but also about my father taking me to an event where he and I would be seen by my friends. It may have been the first time.

By the time we arrived at the Civic Auditorium, the circus was underway. We took our seats and were immediately caught up in the excitement. After several acts, Carl Strong, the ringmaster, introduced the act for which everyone was waiting:

"Miss Rietta will now perform one of the most dangerous and hazardous feats ever performed by any high sway pole artist without the aid of any safety device whatsoever. Let all eyes be upon Miss Rietta."

This was not the petite woman that I had expected but the full woman that I had hoped for: Miss Rietta in a skintight costume, her legs exposed, her breasts buoyant, her lips moistened. This, I thought, would enliven my father; but he sat sullen, his face pointing toward the filth at his feet. I turned back to Miss Rietta.

Miss Rietta was the wife of Arthur Grotofent, Karl Wallenda's half brother. Arthur Grotofent and Miss Rietta had performed with the Flying Wallendas but were now on their own. This was her first appearance in Omaha.

At the top of the pole, fifty feet above the ground, Miss Rietta performed handstands and other acrobatic feats while the pole and audience swayed back and forth. In the middle of this, one of the kids yelled "Doctor Dan," pointing to the other side of the arena. To our amazement, Dr. Dan Miller, a bookish physician and a friend of our parents, was there, dressed in a fez and brassiere, and we waved to him.

With no response from Doctor Dan, we turned back to Miss Rietta, who at that very moment sprung from the top of the pole.

A nine-year-old may be told that humans are confined by the laws of nature or society, may repeat those laws with sincerity, but it is a pretense, a concession to the ambient consensus of which the child at that age is becoming too well aware. But that night we were at the circus, and Miss Rietta could fly.

With her flight all was quiet. It was her neck that broke the silence.

My father was the first to speak.

"Very unusual," he whispered to me and my friends.

As to what happened next, I was not there to witness it (Father was leading us out), but this is what appeared in the next day's paper:

A Shriner in a red fez and a roustabout ran over.

They were soon joined by a white-suited clown, who looked grim even through his painted smile.

Police and fellow performers eased the body of Miss Rietta to the stretcher as the ringmaster ordered the band to strike up another number.

Miss Rietta was taken to the hospital but that, I'm afraid, was a formality. The article went on to speculate as to what had happened at the top of the pole: Miss Rietta, some thought, had passed out and was unconscious when she met the ground. I did not believe that; I believed that Miss Rietta, having slipped, knew she was going to die and wanted to leave us with that last illusion—that we were more splendid, if only for a moment, than the forces which confined us.

Shortly after Miss Rietta my grandfather died. I was leaving school when I ran into Eleanor Slosburg, who was picking up her sons. It was Eleanor who told me that Grandfather had died. She did not say anything about holding his hand or walking through the yard or the suits with the suspenders or the flask.

With my favorite people dead it occurred to me that death was the more wonderful place and that these people had sought it out—it was death that was full of life, full of lovely people (Skimmy, my grandfather, Miss Rietta), and that people who were fortunate enough to be there feared being alive again, which is why they did everything possible to avoid coming back.

5.

A S A CHILD I was disinclined to go to bed until very late and not clever enough to figure out what to do, so I would go to my grandmother's room.

Next to the chair where she sat was another chair, and I

would sit there. I knew she was thinking about the past and I would ask her to tell me about her brothers or grandparents or a friend, it really made no difference, and she would turn down the lights and begin to talk. As the youngest and only girl in a large family, she would tell me of how her brothers would take her to New York and Chicago, where she would be introduced to the gamblers, bootleggers, and entertainers who were her brothers' friends.

The man who bounced her on his knee was Nicky Arnstein and the woman who sat at the table with them was Polly Adler, the society madam. These names, which meant nothing to me, were all the more potent because of it: outside of the darkened room, the two of us, the story, they had no content; no one else could possess or interpret them; no death could take them to another place.

One story (confirmed later by my mother and her cousin Phyllis) began when my grandmother's brother, Doc, received a call from the cantor of the synagogue. The cantor's name was Schwartzkin and he had come home to find his wife gone. This would have concerned him less had he not found her wig. An Orthodox woman rarely left the house without her wig.

Doc called my grandmother, who was a friend of Mrs. Schwartzkin, and Doc and my grandmother went immediately to Cantor Schwartzkin's. As Doc sat with the cantor, my grandmother, for reasons that no one could fathom, thought that Mrs. Schwartzkin might still be in the house. Going from room to room and finding nothing, my grandmother turned her efforts toward the basement.

Mrs. Schwartzkin was not down there.

On her way back upstairs, my grandmother noticed the coal bin. She opened the door to the bin and out popped Mrs. Schwartzkin's head.

The police were called.

When they arrived, a detective went down to the coal bin and began to dig through it for the rest of Mrs. Schwartzkin. By the time he was done, he had found nothing. My grandmother said that when the detective, his face blackened from pawing in the coal, walked into the room with the soot-covered head of Mrs. Schwartzkin cradled in his arms, my grandmother's first thought was, "Bess, you is my woman now."

The police solved the mystery of Mrs. Schwartzkin:

A problem had developed at the Schwartzkin residence that required a plumber. One of the Schwartzkins phoned the plumber. By the time the plumber arrived, the cantor had left for work.

The plumber, having identified the difficulty, set about to fix it. Behind him was Mrs. Schwartzkin, who followed him from sink to drain to pipe and back again, all the while offering her views on the quality of his work. With his work finished, the plumber presented Mrs. Schwartzkin with the bill.

Examining the bill, she claimed that he was swindling her. The plumber had evidently tired of listening to Mrs. Schwartzkin, for he took a wrench from his box and hit her over the head. Fearing the consequences of having killed Mrs. Schwartzkin, he cut her up and distributed her throughout the house.

One night, I was up late and noticed a light on in the family room, which was strange since my grandmother was long since asleep and no one else would have been up at that hour.

Arriving at the room, I looked inside. Sitting on the couch where my father read his books were two men. The one nearest me was wearing a red silk smoking jacket and black pants and was holding a cigarette. The other one, removed by only a few feet, was wearing a red silk smoking jacket and black pants and was holding a cigarette.

Something that night told them to put down their cigarettes.

They rose, brought me into the room, and placed me in a chair. As I sat there, they looked into my eyes.

As they drew closer to my face, I recognized them both as my father.

The next day the doctor informed my parents that my eyes were failing to fuse the views that were appearing separately to my right and left eye. Several days later, my father took the lid off a dark blue box with silver lettering. Inside were a pair of glasses and a booklet. On facing pages of the booklet were identical photographs; the glasses forced the photographs into a single three-dimensional image. I spent a year or more working with those pictures.

The standard view of how sight works is that the images seen by the right and left eyes are brought together in a "final common pathway" that leads to the brain. The pathway of the visual system serves the same function as the motor pathways of the spinal cord.

Recently two researchers, Purvis and Lotto, have uncovered experiments performed over a hundred years ago by a Charles Sherrington that seem to disprove this view.

Sherrington knew that when the human eye is exposed to a light that turns on and off at a certain speed (60 hertz), the eye ceases to see the flickering and instead perceives a continuous stream of light (the "critical flicker-fusion frequency"). If the images from the two eyes were combined in a common pathway, Sherrington reasoned, he could cut the critical flicker-fusion frequency in half by simultaneously exposing one eye to light and one eye to darkness. Sherrington performed the experiment but the results were not what he had anticipated—the critical flicker-fusion frequency stayed the same.

From Sherrington's experiment, Purvis and Lotto concluded that the views of the right and left eyes are fused in the brain and that this fusion occurs when certain aspects of what is being seen suggest to the brain, based upon preexisting information, that the eye is looking at different views of the same object rather than views of different objects. If, however, the latter occurs, a "rivalry" between the views is created in the brain and two images appear to the viewer.

The glasses that I wore as I sat on the couch next to my father turned two Venetian palaces into one; joined mountains and rivers; brought a person who had wandered from himself back inside. It took a while before this happened, but if I sat there long enough the fusion would occur.

After a little more than a year, the doctor announced that the disease had been cured. This he attributed to the effort that I had put into my exercises, and this I attributed to my father's insistence that I do them. Each night, as we left the dinner table, he made certain that I sat next to him and did my exercises.

During my time with the stereoscope, it occurred to me that if my mind was capable of convincing my eyes that two were one, it could achieve the further, more profound, reduction of one into nothing. With that I began to erase myself: whatever was distinct or likely to draw attention, anything that would cause someone to object to me, to like me, to become aroused, to send me to my room or keep me after class, was uprooted and the ground beneath it poisoned; with each erasure I felt stronger and more transparent, and one day I was able to walk unseen back into my own life.

During this period, I would stay up late and every once in a while cross paths with the men in the smoking jackets. While everyone else was fused into one, my father remained two. Like Plato's forms, my father had one imperfect existence, which walked around our house, went to the country club, ignored his children, and ran an optical factory, and one ideal form, which was invisible to everyone except his son, who, by the failing of his eyes, was able to see him. In this ideal form, beyond frailty and fear, he looked out upon the rest of us. It was this transcendence that I sought as a child, not caring what would lift me or had lifted him to it, whether death or whorehouses or other fierce propellants.

6.

IF I WANTED to walk through the neighborhood in which we lived, I would either cut through the Grosses' yard or walk down Sixty-ninth Street and turn left onto Underwood, which wound through the center of Fairacres and was shaded by elm trees. The city of Omaha was set up on a strict grid, so Fairacres, with its curved streets, was unusual.

I have had the advantage of very smart brothers with a wide range of interests. Harlan and Bruce shared an early interest in architecture. By the accident of being their brother, I was introduced to the writings of Frederick Law Olmsted. From these I learned that if there was an immediate precedent for the suburb in which I grew up, it was in Olmsted's design for Riverside, Illinois. That plan included curved roads, yarded houses, open spaces, and an absence of industry that would define the American suburb.

There is a tendency among some to view Riverside as the mature expression of social changes underway since the eighteenth century, changes effected by the self-sought isolation of the bourgeois family. But in my reading of Olmsted, he seemed less concerned with the bourgeois family than with the mental health of each individual, regardless of class.

Olmsted wrote, to take one example from his writings, that "the enjoyment of scenery employs the mind without fatigue and yet exercises it, tranquilizes it and yet enlivens it; and thus, through the influence of mind over body, gives the effect of refreshing rest and reinvigoration of the whole system." As to Riverside, Olmsted explained that the idea was "to suggest and imply leisure, contemplativeness, and happy tranquillity."

Olmsted's views on the influence of landscape on the mental condition of the individual represented a shift in thinking about mental health:

After the French Revolution, Philippe Pinel, who had taken over responsibility for the asylums of Salpêtrière and Bicêtre, observed that by treating the patients in a compassionate way, freed from the restraints and isolation which were the convention, the patients showed improvement. Pinel reasoned that if a change in the inmates' environment could cure the patient, then it followed that the illness itself might be caused by external conditions.

In keeping with the new theories of mental illness, certain physicians proposed that madness was the result of a wearing away of the nerves, causing false signals to be sent to the brain. This, in turn, led to hallucinations, paranoia, and other symptoms of insanity. Physicians identified the competition, crowding, and pollution of the industrial city as a source of the problem. The solution for those who were afflicted by such fatigue was the pastoral setting of the asylum. While the nervous system healed, the mind would be distracted from its "morbid wanderings" by outdoor activity, social events, and reading.

English and American reformers sought to establish asylums ("moral treatment" hospitals) based on Pinel's theories. In 1818, a group of socially prominent Bostonians set up such a hospital. Patients walked landscaped gardens, played games, planted flowers, and participated in social events. Meals were served in dining rooms, and residents dressed nicely. The asylum received a great deal of attention in the press and was regularly visited by tourists, including Charles Dickens. In New York, "moral treatment" hospitals included Bloomingdale Asylum and the asylum on Blackwell's Island.

In 1871, Olmsted collaborated with H. H. Richardson on the design for the New York State Asylum for the Insane. A year later, Olmsted began work on the grounds of the McLean Asylum in Massachusetts and, twenty years later, contributed to the design of the Bloomingdale Asylum.

There is also the matter of Olmsted's own state of mind. As early as 1865, Olmsted's journals refer to his depression, and in 1873, according to one of Olmsted's biographers, he suffered a depression so disabling that he was unable to read for months. In 1895, he was placed in a sanatorium. Years later, he was transferred to the McLean Asylum, where, afflicted with full-blown dementia, he died. Nietzsche spent his last years sitting in a chair at his sister's villa, staring out at the countryside, and mumbling, "I am dead because I am stupid." Olmsted, at least, had the pleasure of looking at a landscape that he himself had designed.

For Olmsted, I suspect, the suburb was a modified asylum —a place where members of society (all members of society, Olmsted hoped) could find respite from the psychological consequences of capitalism. Tranquillity and contemplativeness in a well-ordered setting.

7.

WESTERN HILLS Elementary School was within walking distance of our house. The school was a one-story brick structure with a large playground. The kids in the school were the sort of kids that one finds in a suburban neighborhood. I knew some of them before entering the school. Some I met later.

I did not notice David until third grade, when one morning he announced his intention of bringing a hamster to school but got up in front of the class the next day without it. As he began to describe the hamster, the teacher interrupted to ask him why he hadn't brought it.

David responded, "I've brought my hamster but he's still inside me." David lifted up his shirt and pushed out his belly. He did this with such sincerity that everyone, including the teacher, expected the hamster to pop out. Everyone laughed.

As I grew up, David became known for his tricks. Before the bell for class, he would open the door of his locker, push the front half of his body inside, and then cross his arms so that his hands draped over the back of his shoulders. This gave the impression that David was in the embrace of someone who was standing inside the locker. Teachers, having come to find David, would assume that he had pushed a girl into the locker. Distressed by the sight, the teacher would call out his name. David would swing around and the girl would vanish. At times David would accompany this with the sounds of a sexually excited woman, which was odd since none of us had ever heard the noise of a sexually excited woman.

The last time I saw David was years after we'd graduated from high school. He had taken a room at a local hospital, having performed a stunt by which he made an incision around his neck and then lifted the skin off his head.

If you are a small child and are having sex with your mother, guilt inside you congeals into something thick and plastic and as you move about in life it takes on a variety of forms, including animals and aroused lovers. Pulling the skin off your head allows the hamsters and lovers to get out and makes certain that when they do no one will recognize you.

Another friend from elementary school was named Steve. I remember him as someone who sat alone at a table in the library.

One day I walked over and sat down next to him.

After a few minutes he got up from his seat and asked me to follow him. On the other side of the room was a shelf labeled Fiction and that is where he stopped. Pointing to the shelf, he murmured, "Stay away."

Steve's father owned a store. Steve's father was working there one day when a man came in and shot him; the man left with the money that Steve's father had earned that day.

The world on which Steve stood was so brittle that he did not need the additional weight of someone else's imagination.

8.

WHEN IT became known that my father had ulcers, it changed how people viewed him. We were to do nothing to upset him. One neighbor told me that he was "full of turmoil," though to me he appeared happy.

As to all of this, my father said nothing.

Though he disliked sickness and doctors, he respected the judgment of Dr. Dan Miller (the same Doctor Dan from the incident with Miss Rietta), who, in treating my father's bleeding, suggested that he take up a hobby that allowed him to express himself, thereby relieving the anxiety that had gathered inside him.

My mother believes that this is when my father began to paint.

The notion that anxiety causes ulcers has now been overthrown. The real cause of ulcers is the bacterium *Helicobacter pylori*. The false attribution of ulcers to anxiety led my father to art classes with a group of bohemians, where, beyond the eye of my mother, he was able to smoke and drink, both of which are currently thought to worsen ulcers.

The group he met while painting and sculpting included Ree Schonlau, Sandy Mathews, Tom Bartek, Father Lee Lubbers, Roger Durand, Betty Cutler, and others who congregated at various locations among the decaying nineteenth-century warehouses (known as the Market) near the shore of the Missouri River. Some of these he knew, including his wife's cousin, Eddie Milder. Before he was sent to prison, Eddie and his wife, Cece, were the center of the art scene. Sandy Mathews described my father from those days as "a person who didn't land anywhere for very long."

While all of them would remain committed to the arts, Ree Schonlau would go on to found an art colony at the edge of the Market. She was born and raised in South Omaha, and the art colony in the Market was her attempt to extend the communal experience that she had felt in South Omaha to the abandoned downtown of Omaha.

9.

Having reached an appropriate age, I went off to summer camp.

When I mentioned to my classmates the camp that had

been chosen for me, not one of them had heard of it. My mother told me that I had nothing to worry about since the director of the camp was someone whom my father had known for many years.

A month later, the director of the camp appeared before me on the stage of a small lodge in upstate Wisconsin. He was nothing like the tall woodsman whom I had expected. He was short, middle-aged, and not particularly muscular.

The director introduced himself to the campers by telling a story—the story of a boy who could not find his place in the world. The boy had searched and searched, until he discovered that he had a talent that no one else had—tap dancing on roller skates. So capable was he at this that he found that with relatively little effort he could tap-dance on roller skates up and down a staircase. If this were not enough, he was soon tap-dancing on roller skates up and down a staircase with a girl on each shoulder. At the outbreak of war, the United States asked him to perform in front of the troops, the message being, "If this young man can make it up the stairs with a woman on each shoulder, you can whip the Nazis."

With the evening growing later and colder and the story veering from its end, I was thinking about how to get back to Omaha, when the lights in the lodge went out. We sat in the blackness for several minutes. There was no word from the director.

The lights came back on, and standing there in front of us was our middle-aged director, a staircase, and a pair of roller skates. Up and down he went, humming a tune, accompanied by the percussion of his own feet.

From such a man one would expect an exotic selection of counselors, and in no way did he disappoint. One of the counselors was a man with whom I spent a fair amount of time but whose face I never saw, owing to his habit of wearing a gray veil. This, he confessed to me, was the result of his long-standing fear of being bitten in the head.

This man taught me golfing and shooting, and though he was not an expert at either, he was an expert at both, by which I mean he had perfected the art of shooting golf balls. He and I would go down to the road leading to the camp with a gun, a club, and a bucket of balls. Once we were set up, he would—from behind his veil—shout "Stroke," I would club the ball, and he would shoot it. After fifteen minutes of this, we would trade places.

This gentleman had a friend, and the two of them would play cards in the evening. The friend was skilled at playing cards and one day asked me if I would like to learn to play. Soon I was playing with the two of them and a couple of others who comprised the regular game, which they held around an old picnic table.

These same men would occasionally take me into town for a meal and a beer. The drive was through a dense crop of trees and they would point out sights of interest in the forest. One of the men pointed into the woods and asked me if I saw a light. "That's the motel where I abated the maidenhood of Loraine Johnston."

There was also a counselor who spent his afternoons sleeping on a lawn chair in the middle of the dirt. He wore seersucker jackets and button-down shirts, and had a reddish goatee. He told us that he would answer any of our questions but first we had to convince him that the camp existed. He was Pyrrho of Elis for every argument that we threw out—"If lunch is an illusion, why do we feel so bad after eating it?"—he would counter with an observation which threw everything back into doubt.

Toward the end of camp, the skilled card player gave me a warning: he told me to never forget that there would always be someone who was a better player. What followed was a series of lessons, assisted by his companions, on how to beat a better card player. The instruction here involved dealing from the bottom of the deck, making use of an accomplice, and reading a marked deck (which he happened to have).

The key to all of this was not the hand but the eye.

According to my instructor, there was no hand so fast that a person could not follow it. What allowed the hand to execute the trick, as the greatest conjurers knew, was the seemingly inadvertent glance: the attention of all audiences will shift from the hands and follow the glance of the conjurer, freeing the conjurer to do the trick. *Avoir l'oeil* — to have the eye — is the term for it.

This was a lesson I had already learned from my father. Desperate for the visible, people were willing to watch his earthly, imperfect self, freeing him to reside elsewhere.

While the camp set aside a weekend for parents, there was no question of Norman Rips showing up. His dislike of hiking, fishing, large and small animals, sleeping outside, and the outdoors was well known.

The closest he would come to Wisconsin was Chicago, and it was there, at the end of the summer, that Harlan and I would meet him. I was quite looking forward to this, given that I'd never been to such a large city, and also because my grandmother, who was a great baseball fan, had assured me that I would be attending a baseball game. Before boarding the bus, Harlan and I received a note informing us that our parents had changed plans and that instead of Chicago, we would meet them in Rochester, Minnesota.

If I seemed dejected upon arriving in Rochester, my disappointment was assuaged. Rochester, my brother Harlan and I were informed, was the site of two great attractions, both of which we would visit. Another excitement came when opening my suitcase, in which I found a marked deck of cards, with a note reading: "One day you'll need it." I practiced with the deck for the rest of the trip.

Midwesterners, most especially myself, suffer from the suspicion that whatever they have is available elsewhere in preferable form. Hence the tendency to attach exaggerated significance to those few institutions that are unique. Rochester's Mayo Clinic is one. For as long as I could remember, the Mayo Clinic was the place to which my relatives were taken as soon as local remedies failed. To my young mind, there was the air of the miraculous to the place. The Mayo Clinic was the cause of my excitement but also of my concern—an acknowledgment that we must be there because someone was sick. The person who was sick was my father.

When I asked Mother what was wrong with him, she said they were not entirely certain and she would tell me as soon as the doctors were through with him. All of this came as a surprise to me, since Father seemed healthy and I'd heard nothing of an illness prior to our arrival in Rochester. Other than the ulcers, he was slim, strong, and he complained of nothing. The exception to this was that he smoked. Having done so since he was a child, he still had a two- or three-pack habit.

Mother was one of the first whom I knew to warn people of the effects of cigarette smoking, and she was constantly pressing Father to give it up. Father would tell her that he was cutting down and then would smoke where he would not be discovered. This usually took place in the basement or in the optical factory.

After a day touring the Mayo Clinic, and with my father still under inspection, Harlan and I went off to Rochester's second-greatest attraction: a cream-style corn factory. The factory was much larger than any factory that I had seen before. As we entered, we were told that it was one of the corporation's largest and most modern facilities.

We began on a catwalk, which extended above the area where great stacks of vegetables were washed, scrubbed, and separated from their husks. There was a beauty to the repetitive motions of the workers and their machines, and unlike the quiet of the hospital, there was noise here, a comforting bellow that was the song of those machines. I now saw my future: it was there among the workers and machines and goods that were sent out to the mouths of America.

Upon entering the next room, I saw hundreds of machines spewing thick balls of yellowish liquid into metal cans, one ball so rapidly after the other that they appeared as a continuous stream. Many in our party had to turn their heads. I began to think about another career.

Back in Omaha, I returned to sitting around the house, visiting with whoever happened to be around. One of my favorites was Roy.

None of the garbage in our house was ever thrown out. Instead, it was fed into a large and antiquated incinerator that occupied a corner of the basement.

Roy was in charge of the incinerator. Once an employee at my father's factory, Roy was a giant black man with a deep voice. Father trusted no one with the fire but Roy, and every week Roy would arrive at our house, driving a red pickup truck and wearing coveralls. He would let himself into the house, walk down the hallway, say hello, and then descend into the basement. Fond of Roy, I would follow him.

Though I attempted to be inconspicuous, Roy always knew that I was behind him. Roy knew this because he walked down the stairs backwards. This was not limited to our house; Roy walked down all stairs backwards.

Admiring Roy and knowing that he would not walk down the stairs in such a way unless he had good reason, I began to walk down stairs backwards.

One of the teachers at school asked me about this and I told her that I did not know the reason but that I would ask Roy. Later that week, as the two of us were walking backwards down the steps, I asked him. As we reached the bottom of the steps, Roy sat me down on a picnic bench that had been brought in for the winter and asked me whether I had ever heard of anyone tripping down a flight of steps and falling to his death.

I told him that yes, I had heard of this.

"And when those people fall," Roy continued, "do they fall forwards or backwards?"

Forwards, I was quick to answer.

Though it should now have been obvious to me, Roy completed the logic: "If you walk down the stairs backwards and trip, you're still on the stairs. The other way, you end up dead at the bottom of the steps."

We made our way to the incinerator.

A chute extended from the first floor of the house to the basement, and beneath the chute was a garbage can, roughly four feet high and three feet in circumference.

When the can filled up, another can was moved into its place. As Roy shoveled garbage into the incinerator, I told him about camp and the Mayo Clinic. I also told him that I was worried that Father had cancer.

I could tell from Roy's face that he knew exactly what was going on with my father.

Roy reached down into the pocket of his coveralls and pulled out a shoe. He then opened the hatch to the incinerator.

"Throw it in, boy."

I took the shoe and tossed it in.

"What's wrong with your father isn't cancer; it's something else."

What?

"His toes."

One night while preparing for bed, my mother noticed that the tips of my father's feet were not the sweet peach color that marks a healthy toe but rather a billiard-table green.

The green would not come off and local doctors were consulted. With the green spreading, the doctors admitted that they had no idea what was taking over my father's toes, and a trip to Mayo's was scheduled.

After innumerable tests, the conclusion was reached that what my father had was not only rare but had never appeared outside of equatorial Africa, which was even more mystifying since my father had never been to equatorial Africa, or even Africa. The danger presented by the disease—the rapid spreading of the green up his legs, thighs, and beyond—required immediate action. The tops of his toes were sliced off.

But that was not the end of it. This disease not only ate flesh but could live for extended periods of time in inorganic material. My mother was instructed to burn my father's shoes and socks. This is where Roy came in: prior to my discussion with Roy as to the logic of walking backwards down stairs, he had been asked by my mother to carry out the burning of my father's shoes, socks, and slippers.

10.

AT THE TIME I entered Lewis & Clark Junior High School, the state was on the verge of its centennial, and one of our teachers announced that the entire semester would be spent discussing Nebraska. This teacher was a fellow of some learning, and it was not clear how it was that he came to be teaching a class of thirteen-year-olds. Whatever the reason, he was most certainly not pleased with where he was, because he used almost every opportunity to demonstrate his contempt for us. Since I was well on my way to transparency, this fellow's quips were of little concern, but others felt differently. Because the teacher had a strong and rubbery mind, there was little hope of retaliation.

Not long after we had begun our study of Nebraska, the teacher decided to test us with a surprise question. What, he asked, was the biggest problem facing the State of Nebraska? None of us knew the answer, but this did not stop the teacher from calling on us to offer up our small thoughts, each of which was dismissed with some form of ridicule.

Having gone through nearly everyone in the class, he came to a name that none of us recognized. The teacher repeated the question and waited for the answer. The young man took his time, his head cocked slightly upward. Then, with all of us waiting, he announced:

"Vaginismus."

The teacher stepped backward. None of us knew what that word meant, but from the entanglement on our teacher's face, we knew that he did. The young man would become my life-long friend.

Sex was not a subject of discussion at our house. Four boys and I cannot remember a lewd remark, a joke about sex, a pornographic book or magazine. My brothers had girlfriends but they rarely came over to the house.

There was one exception: my mother's attempt to provide her sons with sex education, an idea that was just being introduced to the Midwest. My mother's efforts here were modest: when each of us reached the age of fifteen, a book appeared in our bedroom. The volume was a primer on sex in the animal world, beginning with protozoa, making its way up through pulpy vertebrates, and ending with pachyderms.

Returning from college after his sophomore year, my brother Harlan announced over dinner with the family that a woman had introduced him to a sexual position in which the man lies face-to-face on top of the woman. She recommended this as a possible alternative to approaching a woman from behind and draping one arm over her shoulder.

On that same trip back, Harlan reported that he had met a gentleman who suffered the embarrassment of having a "merkin" inadvertently affixed to his forehead, the result of an unspecified encounter with a woman. According to the story, the man had walked around town for a couple hours before realizing that it was up there.

This story was more disturbing than the first one, because when I went to the dictionary to find merkin, the closest I could come was *merchanti,* which was defined as a "bazaar merchant" or "profiteer." It made no sense that he would have either of these on his forehead, so I was relieved to know that what was actually up there was, as my brother defined it, a "toupee for the vagina."

One day that same summer, the summer of the merkin, my father and I approached a man lying on the street just east of the Civic Auditorium. He was a Native American.

Growing up, there were many Native Americans in Omaha, and Father told me that they had been more numerous when he was a kid. Sol Parsow, the clothier, also told me of Native Americans walking on the street in front of his store in downtown Omaha.

The Omaha tribe, as my father informed me, believed in a power that was inaccessible to members of the tribe. Fletcher and LaFlesche define the Omaha term *Wakondagi* as a power that permeates all consciousness but which remains mysterious—a concept similar to the description that the Bearded Priest provided of the Other as found in the writings of Levinas.

In 1937, Rudolf Otto published *The Idea of the Holy*, in which he identifies the "numinous" as a critical part of religious experience. The numinous, according to Otto, is the experience of "inescapable and irreducible mystery." Or, to put it in another way, the experience of religion untouched by the alloys of rational or moral meaning.

The effect of the numinous, according to Otto and Levinas, is piety—a sense of smallness and unworthiness ("I that am dust and ashes," Job 42) in the face of that which we cannot know. This, too, is found in the mythology of the Omaha tribe, whose prayers are characterized by humility and the sense of insufficiency that comes from longing for something that cannot be possessed.

The man, the Native American toward whom we had been walking, was now directly before us. His eyes, which had been closed, were open. They were vital and drawn upward, aligned with an arc that extended out of the Civic Auditorium, the arc that had once held the flight of Miss Rietta.

THE FACTORY

I.

THE FAILINGS of my eyes had made me curious enough about the process of seeing that when my father suggested that I come to work at his factory during the summer, I agreed.

Each morning at seven-thirty I was delivered into the hands of Siegfried Christianson, the manager of the surfacing room. If my grandfather and father had searched to find the most intimidating man to manage their optical factory, they could not have done better than Siegfried Christianson, who was both taller and broader than any other man on the floor. With his light hair, large jaw, and powerful arms and fists, Siegfried ("Fred") inspired obedience. Even if one were inclined to protest, it would have done no good: Fred had the confidence and authority of my family, and for that reason there was no appeal from his orders.

In the middle of the twentieth century, the optical industry was little different than it had been for a hundred years: lenses were ground and glasses assembled by local laboratories, with each laboratory comprising a small number of highly skilled men.

My father had no interest in this system. As soon as he took over the optical factory, he began experimenting with machines that could mass-produce lenses, and when he was satisfied that those machines could do the job, he advertised across the country, offering prices that were the lowest in the nation, and as a result he received orders from doctors in nearly every city in America.

For the system to work, he needed to produce an enormous number of glasses quickly. Fred, with his authoritarian manner and his mastery of making glasses, was able to help my father achieve this.

During that summer, Fred moved me from one job to the next, always making certain that my incompetence would in no way interfere with the pace of production. I, on the other hand, spent my days avoiding Fred, my assumption being that a man like Fred had a temper and if provoked would unleash it.

When an order first came into the plant, the prescription was written down on a sheet of paper, placed in a box, and then sent up to the room where the lenses were kept. Lenses were selected from inventory based on the size of the "correction" (the term used for the prescription), the dimensions of the frame, and the color of the glass. The lenses were placed in the box with the order and the box was sent off to a room where plugs of metal were affixed to the back of the lens. The box was then delivered to Fred Christianson and the surfacing room.

The first stop in the surfacing room was the generators. Each generator had a long arm with a clamp on the end. The clamp held the lens. The lens was then dragged across a diamond-coated wheel spinning at high speed, which produced the desired curvature.

The heat created by the wheel was so great that if the lens was not cooled it would explode. For this reason, a continuous stream of liquid was applied to the lens. The liquid rained on the operator, and the noise of the machines was something close to the noise that I imagine a cat makes when it is fixed.

Because the arm of the generator was four or five feet off the ground and heavy, it was assumed that the machine could only be operated by a tall man of some strength. But standing next to me that summer was a woman who was shorter than any man in the plant.

When she first came to the factory, she was assigned to a clerical position. Intrigued by the generators, she approached my father. No woman had ever worked on the generators. My father looked at her and then the generator. She was shorter than the arm.

Without a word, he walked over to a box that was sitting on the floor and dragged it in front of the generator. She got up on that box and stayed on top of it for another thirty years. Among lens grinders, there were few better.

Once the lenses were correctly cut and polished, they were sent on to be fitted into a frame. This was usually accomplished by aid of a pattern that was the precise shape of the frame. A machine cut the piece of glass to match the pattern. In some cases, where the frames were unusually shaped or the lenses too cumbersome to fit into the machine, the lens would be cut by hand. There were two men who did this, and they were located in the back corner of the surfacing room.

One of these men was named Charlie, and it was Charlie to whom I was assigned.

When Charlie was not cutting lenses, he would wander around the factory and talk to people. He told me that he was raised on a ranch. He had the polite manner that I identify with people who are from the countryside of Nebraska. Despite living and working in Omaha, he continued to wear his cowboy boots, cowboy shirt, and a belt with a silver buckle.

Charlie and I would have lunch together and every once in a while we would have a drink after work. There was a bar across from the plant, and it filled up with workers when the plant closed.

There was a jukebox in the bar and not infrequently people danced. Workers, who flirted with each other in the factory, would meet and dance at the bar.

So that I would be able to join in with the others, Charlie offered to teach me the two-step. There was no doubt that Charlie was a good dancer. No woman turned him down, and he seemed to be as good as if not better than the other dancers. There was even a point in every dance when Charlie, holding the woman close, would stop, separate himself from the woman by taking a few steps back, and then bend the top of his body downward in a modified bow. This gesture always caused a giggle from the woman.

Charlie's lessons did not end with the two-step. He taught me how to approach a woman, the proper way to ask her to dance, the etiquette of offering a lady a refreshment. Charlie knew the formalities of love, and he gave me the education that I had not received at home.

Near the end of the summer, I was still working with Charlie when a lens that I was cutting slipped from my hands and rolled under the bench. This was not uncommon, particularly for someone who was still learning. With Charlie standing at my side, I bent down to get the lens.

Reaching for the lens, I noticed something else under the bench. Bending closer, trying to focus, my nostrils came dangerously close to an uncircumcised part of Charlie.

The incident was so curious that on my final day of work, I was moved to ask Charlie about what had happened.

"Charlie," I began, "the other week when I was under the bench, I saw something—"

"A penis?" interjected Charlie.

"Yes, Charlie, a penis."

"Don't worry," Charlie comforted me, "it wasn't mine."

"Charlie," I said, "if you don't mind my inquiring, whose penis was it?"

Charlie thought, and then, moving his hand to the inside of his crotch, gave a tug on his pant leg. I looked down toward his boot and there it was again. But as Charlie had said, it was not his.

The object was an artificial penis that was attached to Charlie's cowboy boot. With Charlie standing straight, no part of the phallus was visible, for his pants covered it. But when he bent his leg or lifted his pant or bowed, the tip would peak out from under his hem. This was the part of Charlie's two-step that I had missed: when bowing to his partner, he would glance downward, making certain that his partner's attention found its way to the bottom of his leg.

On days when I was not having lunch with Charlie, I would go upstairs to the fusing room. There, lenses ground with the correction needed for reading were set on top of those used for seeing at a distance; the two lenses were then placed on a conveyor belt and rolled slowly through a brick oven. The oven reached extraordinary temperatures, thereby joining the two lenses into a bifocal. If the conveyor changed speed, or the lenses were jostled or even exploded owing to flaws in the glass, the entire batch of lenses would be ruined. For this reason it was necessary for someone to be in the room at all times watching the oven, and there were very few who could withstand the heat of that room. A cousin of my father's, who also owned an optical factory, worked the fusing room naked.

This room was the province of my grandfather and the men who worked under him. With this, neither my father nor my grandmother interfered. As a young man, my grandfather was in a position to retire. Instead, he returned every day to the fusing room, and did so well into his eighties. He was indistinguishable from the other workers: he wore the same work clothes, worked the same hours, and ate with them.

Some of these men had been in concentration camps, and standing next to the oven, staring into the fire, the numbers on their arms glowed.

Returning from lunch with my grandfather, I was impressed by the difference between the men who worked with my grandfather fusing lenses and those who worked in the surfacing room. The former were easily categorized (central and Eastern Europeans, many of them Jews); the surfacing room was a more puzzling assemblage.

Charlie knew a lot of the people in the plant, and standing in the corner we were able to see everyone on the floor. He would point out details of the people he knew: the deaf and the cripples, the criminals, the alcoholics, the ones who were sexually available, the ones who had been in mental institutions. At first I dismissed this as the imaginings of Charlie, but the longer I worked in the plant, the less likely I was to dismiss his observations; he was a sociologist, albeit one who had a penis strapped to his shoe.

The way that hiring worked in the surfacing room was that Fred Christianson would review applications, select those people he wanted to interview, and then present the ones he had chosen to my father, who would make the final decision. Fred's decisions were based on theories that other employers in Omaha might consider unusual: Fred hired the deaf because they would not be distracted by what was going on around them; obese people were more likely to stay in their seats; men on parole were less likely to leave town. As my brother Harlan has pointed out, if a person whom Fred hired—a cripple, for example—worked out, then every person Fred hired after that would be a cripple until one of them failed to do their job or a flaw was revealed in Fred's theory (the efficiency that was gained by using deaf people, for example, was lost when they used their hands to talk to other deaf people).

As Fred Christianson grew older and his decisions as to hiring became more eccentric, there were certain people in the company who came to the opinion that he was no longer able to handle his job. One of these, an assistant manager, a young man, approached my father to say that the time had come to replace Fred and that he was ready to take on the job. The young man did not, it seemed, care that Fred was standing close enough to overhear the conversation. My father listened politely, looked at Fred, and then turned back to the young man. As Fred tells the story, he saw in my father's eye "a look that meant he was about to punch the kid out, so I grabbed your dad and pulled him away before he struck him."

Fred would not have been able to put his theories into effect without the help of my father. As one of the workers told me, "Word around town was that if you couldn't get a job anywhere else, there was a guy downtown, Nick Rips, that would give you a shot. We were a factory of freaks, but we were loyal to Nick." On more than a few occasions, the police would come to the factory to arrest someone on an outstanding warrant who was happily working away at their job. As they were led off, they had one comfort: the knowledge that no matter what their crimes against the rest of the world, they would be welcomed back into the society that had been assembled by my father. Whatever benefits my father conferred on the workers, he received much in return: the people in his factory strengthened him.

One day early on, Nick Rips looked out over his workers and decided that there was still something missing: blacks. With that, he walked around the factory and to each supervisor gave a very specific instruction: hire blacks. The order was carried out and very quickly blacks and, shortly thereafter, Hispanics, mostly women, became a significant part of the workforce.

The hiring of women was itself unusual: historically, optical shops were closed to women. Nick Rips made certain that women were hired in his factory, and over the years the factory came to be dominated by them. Norman Zevetz, who worked in the factory for over four decades, said that "what was important was not that blacks and women were hired but that your father listened to them. He was curious about their lives. He talked to them and they talked to him."

My father hated sitting in his office. He wanted to be out in the factory, among the workers. But he also worked: if someone was ill or the work was backing up or there was something that needed to be fixed, Father was the one who stepped in. On countless occasions I would see him working the generators or polishers, hand-cutting lenses, or writing up orders; there was a small retail shop on the first floor of the factory, and when time allowed he would sit down there and fit people with glasses.

Before putting lenses into their frames, each lens is examined for imperfections (minute spots that were missed by the polishers, nicks around the edges of the glass, small bubbles inside the lens). If any of these are present, the lens is sent back. Even then it might not be right, and it would have to be sent back again, to be cut and recut, polished and repolished. Every so often a lens would come through that was so clean, so utterly free of everything in this world, that it would vanish and it was in those moments that I came to appreciate the beauty of the invisible.

2.

FRANK WILLIAMS and Reggie believed that the woman in the painting came out of the factory. Such a woman, they reasoned, could be counted on to keep quiet. A model in an art class could not; besides, Frank Williams claimed that there were no black women modeling for art classes at the time my father was painting. If there were, he assured me, he would know about it. I was inclined to believe him.

The person who was most likely to recall the people who were working in the factory was Fred Christianson. Finding him was not so easy. No one had seen him recently, and there was some speculation that he was no longer alive. I was given an old number for Fred, but no one answered.

I went to see Williams.

The parking lot was empty, as it had been on each of my previous visits, but I stood at the door, as I had done on each of my previous visits. After ten minutes, the door opened. It was Williams.

We went into the basement and Williams locked the door. As Williams watched the monitors that were trained on the parking lot, I related my difficulty with finding Christianson.

What, Williams asked me, did Christianson look like?

"He would be in his eighties, tall, Germanic—"

"Tell me no more!" Williams interrupted, turning back to the monitors, as if looking for Christianson.

After a minute or two, I got up to leave.

"Find your father," Williams began, still staring into the screens, a mystical tone to his voice, "and you will find Christianson."

Surely Williams had misspoken. He meant: find Christianson and you will find your father. The odd thing was that Williams said so little and spoke so slowly that it was hard to imagine that he would mix things up.

As I rose to go, I looked past his shoulder. There was nothing on the monitors but images of the empty parking lot. I climbed the stairs to the first floor, making certain to close the door to the basement behind me.

At my car, I stopped. I thought back to Williams in the basement, and to the monitors. My car should have been on one of them. It was not. Williams was watching screens that had nothing to do with the parking lot.

I had been given an address for Fred Christianson and drove to his house. No one answered. There was no one on the street and many of the houses were vacant or boarded up. I returned home and continued to try the number I had been given. One evening a man picked up the phone. It was Fred Christianson.

We talked about my father and the factory.

I asked him about the black woman in the painting.

Fred said that he knew of no such woman and that as far as he knew my father had conducted himself in an entirely proper way with the women in the factory.

In the course of our conversation, I asked Fred whether I had been knocking at the right door. He told me that I had not. He had lived in that house for a very long time but had recently moved.

Fred said that as a young man he had lived on the north side of Omaha (a black neighborhood) because it was there that he could find a house that was affordable. The woman whom Fred married fell ill, and the expenses necessitated by her treatment were large. Though the neighborhood changed, growing (according to Fred) more dangerous, he decided to stay where he was.

One day Fred noticed that there was a girl in the street who had no shoes. She was playing with Fred's daughter. With the consent of the girl's mother, Fred and his wife took the girl in. Soon Fred became known in the neighborhood as someone who was willing to take care of kids who were in trouble. His house began to fill up; some lived with him for a month, others for years. The beds, couches, and basement were always full. Fred treated them all as his children, feeding and clothing them and, when they got older, helping them find jobs. At the same time, Fred continued to take care of his wife.

The police department and halfway houses soon knew of Fred Christianson, and if they had a child who needed special care, they would call him. He rarely declined their requests. As Fred grew older and his wife weakened, Fred had little time or money to maintain his house. By the time Fred's wife died, the house was collapsing and there was very little that Fred could do about it. Whatever he was able to save had gone to his wife and to the kids who were living with him.

Only months before I found Fred, he had received a call from a woman. Fred had raised her. After graduating from high school, she had received technical training and was now working in Omaha. She wanted to invite Fred over for lunch. Fred thanked her but declined the offer. The woman persisted, and a week later Fred arrived at her house.

When he entered the house, which was nicely painted and decorated, the sort of house Fred had dreamt of having, he was greeted by his host. There were a number of other women as well, and as they introduced themselves it became clear to Fred that they were all women whom he had raised. Some of these women had come in from out of town; many had not seen Fred for years.

They all sat down to lunch.

Halfway through the meal, with everything going well, Fred got up to go. He felt uncomfortable with the attention; he was also worried about his daughter, who lived with him and needed to be taken to the doctors twice a day.

As Fred was leaving, the women handed him an envelope. As he walked to his car, he opened the envelope.

Inside was the deed to the house in which they had just had lunch. The women had purchased the house for Fred and paid to have it fixed up.

Today Fred Christianson leaves this house early every morning to deliver bags of groceries and other supplies to the people in the neighborhood who have trouble leaving their houses. During the day, he drives his daughter to the doctors.

There is a small graveyard around the corner from where Fred lives, and as he passes out groceries in the morning, he walks by it. My father is buried in that graveyard.

"Find your father and you will find Christianson."

3.

THE SUMMER before my first year at Central High School, I could not think of what to do. Most of my friends were clever enough to come up with something, but I was not. Consequently I ended up in classes that were offered at the high school.

I missed the first week of class, so as I walked into Central High School there was no sign telling me where to go. I found a security guard and was directed to the basement. As I made my way through the hallways, I thought about all the members of my family who had been through the building—grandparents, parents, aunts and uncles, brothers, cousins—and this was beginning to stir in me a certain emotion, when it struck me that of all of those relatives I could not remember one of them mentioning summer school.

In the basement, with the heat climbing to a level that few could have been expected to endure, it occurred to me that Central High School had no air conditioning. As I reached my class, the entirety of my clothes were filled with perspiration joined by its accompanying odor. Now it all made sense: no one in my family had taken summer classes at Central because no one in my family was quite so stupid.

If the heat made me uncomfortable, it was worse for the others in the class, who, with one exception, were in the summer ROTC program and were consequently dressed in uniform. The one not in uniform was dressed in the nicest of white suits. As I walked into the room, he nodded to those in uniform. They stood. He remained seated.

As I sat, he again nodded. The men in uniform sat.

With the regiment back in their chairs, I got a good look at their commander. He was none other than James Ross III, who had several years before introduced me and my classmates to "vaginismus" and now had taken control of an ROTC unit, which, in the manner of the most skilled despot, he directed with the slightest motions of his head.

The class that day ended at noon, and James Ross III and I decided that instead of taking the bus back home we would walk downtown. The downtown of our parents' generation was gone: businesses had moved west to the suburbs, the retail shops had closed, and there were only a few places to eat. Drunkards were propped against the sides of buildings. Buildings were propped against the sides of drunkards. There were still some prostitutes but most had moved elsewhere. You could stand on the sidewalk for some time before anyone passed.

James remembered that one of his relatives, a cousin on his father's side, was the manager of a downtown hotel. The hotel was within walking distance of Central.

Upon entering the hotel, the New Congress, we were greeted by a man named the Colonel. Jim's relative was not in, but the Colonel invited us to wait in the lobby. The lobby, located two or three steps from where we were standing, included a sofa, two chairs, and a black-and-white television set. We sat on the sofa.

A few minutes later one of the residents of the Congress joined us, and a few minutes later another. Word made its way upstairs that there were newcomers to the hotel, and the residents came down to have a look. These people were a great deal merrier than the Colonel, and after a few hours of chatting and watching television, James and I departed.

Though James's relative, Glen Eckert, never showed up, we so enjoyed ourselves that we returned the next day. This time we stayed longer. Each day we extended our stay, meeting more of the residents. Around two o'clock, most of the residents would come down for cocktails. Since the Congress did not have a bar, the residents brought their own bottles; since the Congress had no glasses, the bottle was passed from person to person.

After a week, Eckert appeared. He had once been a rodeo star, riding under the nickname Dusty Tex, but according to James he "busted up" in Denver and moved to Omaha, where he lived with James's dad. James's father found Eckert the job at the Congress, but it was not the sort of thing Eckert wanted to be doing, so he would disappear for long periods in the middle of the day.

This is where the Colonel came in.

According to the Colonel, Eckert had deputized him to make certain that the hotel was "running as a first-class operation and that there was no fighting in the lobby." If fighting broke out, it was likely to be during cocktails, and Eckert provided the Colonel with a gun, which on no few occasions was brought out by the Colonel to clear the lobby or to chase away prospective guests whom the Colonel did not believe met the standards of the Congress. It was only recently that James told me that he believed that the Colonel was never actually an employee of the Congress but was rather a half-sane man whom Eckert had convinced to work for him without pay.

One of the first residents of the Congress whom I met was a gentleman who had been in the war with Eckert. Eckert liked him because he didn't cause a lot of trouble. I liked him because he told me stories about the war.

Toward the end of the war, the man was stationed on an island in the Pacific. There had been some casualties and he was assigned to bury bodies. The natives had buried their dead in shallow craters scattered around the island, and he decided to follow that custom. One day while he was digging, his feet began to shake and then his legs and, with his entire body wobbling, there was an explosion. Opening his eyes, he was floating above the island at the top of a thick column of bones, some recent, others ancient, and next to him was the man whom he had just put in the ground, but who, by force of the explosion, was upright and appearing to be looking around, taking in the view, which was not such a bad idea, since from up there they were able to see the neighboring islands and parts of their own island that they had not yet had a chance to visit.

What he had not been told when he was given his assignment was that the craters had once been small volcanoes and that some of them were still active. His misfortune was to be shoveling into a volcano at the time it ejaculated its dead.

When he came down, he did so on an assortment of bones and skulls, and it was a shard from one of these that removed his jaw. Missing his jaw, he had an enormous red dewlap, which would whip across his face when he got excited.

His best friend at the Congress was a short, soft-spoken fellow who lived on the second floor. I remember little about the man (though I think his name was Mike), other than the day when, as he was ascending the staircase to his room, something struck him and he turned around, mounted the banister, and allowed gravity to return him to the lobby. The flight did not go as planned, and he ended up inside the television set, which was the sort of commotion which upset the Colonel.

If we were not at the Congress, there were two movie theaters (one pornographic) and a place called the Rocket. The latter was owned by a midget named Maxi, which I am not certain was the name given to him by his parents or an ironic acquaintance, but it fit him, for other than his height he was a big man. The Rocket had a bar and a couple of pool tables, but it was widely known that Maxi, who dressed in a white fedora and a leisure suit (36 short, according to Sol Parsow, who had the clothing store around the corner), was willing to accommodate the gambling interests of his clientele.

The summer that I was hanging around the Congress was the same summer that my brother Lance, who was away at college, decided to spend a few days in town. Whether he noticed that I wasn't doing much, I do not know, but one day he walked into the room where I was sitting and handed me a book on philosophy.

What made sense to me in that book was the section on skepticism, and I took to reading as much as I could find on the subject and in it discovered a certain calm. (The ancients recommended skepticism as a means for attaining complete peace of mind.) I could barely achieve the simplest addition, knew nothing of science, still had not read a word of fiction, but I knew Pyrrho of Elis and had committed to memory Hume's observation that the senses "cannot operate beyond the extent in which they really operate"—the vivid tautology from which philosophy has arguably yet to recover and of which I had long before become convinced by stereopsis and by working in an optical factory.

It was in this book that I first encountered the name of Levinas.

Husserl had attempted to answer the skeptics, and Levinas had written interpretations of Husserl. I paid no attention to Levinas until the Bearded Priest mentioned him over coffee decades later in New York. For him, Levinas's view of the world, quite apart from his commentaries on Husserl and Heidegger, was extremely important. It would take three stories from antiquity and a painting in the basement for me to understand why.

At first I did not know how my father would feel about his son spending his afternoons at the Congress, but it soon became apparent to me that he found the matter of some interest and, for the first time, seemed to pay attention to what I was doing. I would tell him about the man with the dewlap, the Colonel shooting off his gun, the drunks, the prostitutes, the former rodeo star. This was the part of society that engaged my father, and now, in retrospect, I realize that it was for him reminiscent of the Miller Hotel, the family brothel.

Two of his favorite places to dine, King Fong's and the Castle Hotel, were also throwbacks to a time in his life that had passed.

King Fong's is at the top of a long narrow staircase in a building on Sixteenth Street. The interior, which has not changed since the restaurant opened, has thick marble tables, oriental chandeliers, stained glass windows, and private booths in the back. The booths have their own buzzers for calling the waiters. There is a second floor to the restaurant, which during the twenties and thirties was used for drugs and women.

By the time I was working downtown, the second floor was closed but the restaurant was still open. Father and his brother would take a booth, put in their orders with a woman named Yo, and then debate whether it was more efficient to use chopsticks or a fork, with Father taking the side of the fork and his brother arguing on behalf of the chopsticks. The two of them ate there frequently and a lunch did not pass without the argument over the fork and the chopsticks.

The Castle Hotel, not far from King Fong's, was quite near the plant and in its lobby was a restaurant. Helen, the waitress, weighed a couple hundred pounds, dressed in a white frock, and, having shaved her eyebrows, replaced them with two solid arches which ran from below her eyes to within inches of her hairline and then swept back down to her nose.

When I was eating at the Castle with my father, there were few if any customers. This is not as it had once been, for the Castle was once a successful, possibly even swank place to eat. Father's nostalgia, and his realization that if he stopped eating at the Castle, Helen would be unemployed, resulted in his eating there incessantly for years on end, more often than not with no one else in the restaurant.

· · ·

One of the gentlemen in the lobby of the Congress was always reading books. He told me that he had left the town in which he had been living when his house was destroyed by a tornado. He saw it as a sign. Since then he had moved from town to town. Having arrived in Omaha, he ended up at the Congress on the suggestion of someone at the bus station.

The man spent his days in the public library and was full of information, which he was happy to share with me. He told me, for example, that people once believed that if a woman was suspected of being a witch, they would remove her pubic hair because that is where the devil resided. He assured me that it was unlikely, but since I was beginning to go out with women, I might want to keep a watch out.

There was a jewelry store near King Fong's and one afternoon the two of us walked inside. There was no reason for the employees of the store to pay attention to us, since I was fifteen and the man with me was unshaven and his clothes were worn. After several minutes, one of the salesmen asked us whether he could help us (the tone suggesting that he was not at all interested in helping us).

The man from the Congress turned and asked with some condescension: "Do you have black diamonds?"

He may have seen a picture of one of these in a magazine, or even heard mention of it from someone in his family, but such a diamond, any diamond, was so obviously not a part of this man's life that the question made him seem all the sadder.

The salesman replied that he had no black diamonds, and we walked out.

Back at the Congress, the man informed me that though he was enjoying his stay in Omaha, he would be moving to another town that afternoon. I helped him pack, though to be honest he had almost nothing.

Before he left, he stopped in the lobby to say goodbye and to explain to me and the others the mystery of the black diamond: when the pits and fractures that would normally ruin a diamond reach a certain level—reflecting off each other a thousand times—they produce a magnificent stone. I felt proud to know this man.

Years later I would encounter a Belgian diamond dealer who would tell me that this was not at all how black diamonds were created. The man at the Congress, it turned out, knew nothing about black diamonds. What he knew about was the people at the Congress.

THE
SECOND STORY

I.

DOWN THE STREET from where James Ross III and I were going to summer school, Vivian Strong, a black teenager, was shot and killed by a white police officer. Three days of riots followed. The incident occurred not more than a few minutes' drive from my father's factory. This riot was not the first. For several years there were episodic outbursts from the black community, a community that was becoming impatient. The white-owned businesses that had not already left the downtown closed their doors and moved west.

My father stayed.

Shortly after the riots, Father asked me to come to the factory.

He was in the surfacing room talking to Fred when I arrived, so I sat in Father's office. In one of the drawers was his gun and I pulled it out.

When Father appeared at the door, I slipped the gun back into the desk and we headed down the back steps to the parking lot. We drove west and then north. Passing Creighton, we were quickly in that part of Omaha where the riots had taken place. There were no whites.

We turned into a bar that was one of the few buildings that had not been boarded up. Father got out of the car and I followed. We entered the bar, sat down, and Father ordered sandwiches.

On our way back to the factory we passed the public library. The top of the car was down, the air sweet; and there ringing the top floor were the names of Schiller, Goethe, Dante, Tasso, and Corneille, the medallions of Horace, Plato, Seneca, and Herodotus. Sophocles was above the front door and it was to him that Father was pointing.

Father was in the habit of making reference to or even quoting from texts that were, at least to my mind, obscure. The texts were known from a time when, as his friends recall, his interests were in literature and especially the classics.

"Do you know the play *Ajax*?" he asked.

I did not know *Ajax*.

What followed was the story of Ajax—as found in the earliest of Sophocles' known works and as retold by Nick Rips. I was delighted to have this story, for it replaced what would have been the long silence that was the third passenger in almost all of our drives.

His version of Sophocles' story was as follows:

Ajax, the son of the king of Salamis, led the troops from Salamis in the Trojan War. So great a warrior was Ajax that when, in the middle of battle, Athena appeared at his side to give him direction, he refused her help—ordering her to find other Greeks to protect. It was Ajax who fought in single combat against Hector, and Ajax who rescued Achilles' lifeless body from the Trojans, under a storm of arrows.

With Achilles dead, Ajax and Odysseus were the only legitimate contenders for his armor. After the Greek generals awarded the armor to Odysseus, Ajax—jealous of Odysseus and convinced that the decision was fraudulently obtained— set out to slaughter Odysseus and the Greek generals. Before he could carry out his plan, Athena intervened, so confusing Ajax's mind that he mistook the cattle of the Greeks for the Greeks themselves. This figment led him to slaughter some of the cattle immediately, while dragging others into his tent, where he struck and cursed them.

Awakened from his madness, he was overwhelmed by humiliation. ("There he sat, wreckage himself among the corpses, the sheep slaughtered; and in an anguished grip of fist and fingernail he clutched his hair.") To his wife, he declared his intention to kill himself. After she pleaded with him, Ajax assured her that he had changed his mind.

At this point, a messenger arrived with words from a seer. The seer foretold Ajax's death but also explained the cause of his suffering—Ajax's rejection of Athena during the battle with the Trojans. As the messenger finished the seer's prophecy, Ajax planted his sword in the sand and threw himself upon it.

When Teucer, Ajax's brother, attempted to bury Ajax, Agamemnon forbade it: Ajax had committed treason and would be left unburied, to be consumed by animals. Teucer and Agamemnon were on the verge of battle when Odysseus interceded on behalf of Teucer. Ajax was buried and Odysseus was praised.

2.

AFTER MY SUMMER at the Congress, I entered my first full year at Central. At the time, the racial conflicts that had broken out around the high school made their way inside.

By accident, I missed the first outbreak. Father was late in dropping me off, so when I entered the building, the fighting was over. Police were clearing the hallways and a cleaning woman mopped the blood from the lobby.

I knew one of the students who had been stabbed during the riot. Upon recovering, he left Central. Another student was stabbed a couple months later, this time in a classroom. He had said something in the course of a presentation to the class that was insulting to blacks, and on his way back to his desk, a student pushed a pen through his abdomen. He, too, left the school.

The events at school that year caused white parents to talk about pulling their children out of school. For the parents, the choice was a relatively easy one: there were many suburban high schools that had few if any blacks and strong academic programs.

There was some debate in my family as to whether I should be sent to boarding school on the East Coast, with a flurry of catalogs of leafy places and hushed discussions on the merits of "going east." How far these discussions had advanced in our family was unknown to me, but at a certain point, passing me in the hallway of our house, Father said to me, "You're staying," and that was the end of it.

That decision was in line with my own preference. Each morning I would have breakfast at the coffee shop next to the Congress and would often return to the Congress at the end of the day. In this small schedule I found comfort.

The coffee shop and the Congress were under different ownership, and as a result the owners of the coffee shop were on a constant lookout to make certain that the disruptive elements from the Congress, which were not few, remained in the Congress. The Colonel chasing someone through the coffee shop with a gun was the sort of event they sought to avoid. The

Colonel, on the other hand, found the attitude of the owners of the coffee shop insulting and as a result had little to say about them. Certainly he had no interest in being neighborly, though I suspect that he rarely had an interest in being neighborly.

Early one morning, breakfast was being served in the coffee shop and the place was crowded. The owner was walking around, taking orders and pouring coffee, when small drops of water began falling from the ceiling. Excusing himself, he found a bowl and placed it below the dripping. As the bowl filled up, he found a pot. As the pot filled up, he went to find Eckert, for it was obvious that the water was coming from the Congress, which extended over the top of the coffee shop.

While the owner was searching for Eckert, a hand came through the ceiling. Not the hand of a plumber, nor the hand of Eckert. It was a slack, white hand followed shortly by a slack, white arm. A shoulder and then a face. The face of Mike, the resident of the Congress who was constantly infuriating the Colonel, most recently by sailing through the lobby's television.

Several days earlier, Mike had been drinking and decided to take a bath. Having taken off his clothes, he fell dead on the floor before he could climb into the tub. After a day or two the floor gave way, which is when Mike came squirting through the ceiling.

It would have been over quickly had Mike's foot not become caught in the floorboards, causing him to hang upside down over the dining room. This ended when Mike's body separated from his foot. Mike's head, which was soft from a two-day soaking, hit a table and flattened. It was then that the owner returned from his discussions with Eckert.

What the owner discovered in his coffee shop was a small body of water in which floated a pair of Mike's underwear, a danish, and Mike himself, who had on his neck (in place of his head) what looked to be a large waffle topped with a scoop of hair, an item that did not appear on the menu of the coffee shop. Fifteen feet above was a bloated foot and two or three strips of skin rotating in the wind of a fan.

Eckert, having emerged from the Congress, stood at the edge of the water. He was wearing a nice shirt. His hips started to move. He was dancing. It was spring.

3.

I F I SAW little of my father during the week (other than when he picked me up at the Congress), I saw him less on the weekend. During the summer and spring, he would spend his weekends at the club playing golf. During the winter, he would attend college football games or bowl.

The people with whom my father bowled were the workers in the plant. Like other companies, the factory sponsored a bowling team.

One morning I went with my father to the Ranch Bowl.

After bowling, Father and the other members of the team gathered in the restaurant at the bowling alley. I had the Reuben sandwich and, finding it agreeable, decided to come back the next week. From then on I was a regular. Not that I ever bowled. It did not occur to my father or me that I should bowl. Instead I sat in the gallery and watched. Once or twice a game, Father would come over to discuss the mechanics of bowling. I paid attention, and by the end of many, many hours of watching my father and the men from the factory, I came to develop an appreciation of bowling.

In the course of my days as an observer of bowling, I noticed that the game was capable of generating excitement and that this was the one aspect of the game in which my father had no interest. While others cheered or fell to their knees, Father remained impassive.

There was no condescension here. He was a man whose view of himself did not include such outbursts: he would place himself nearby — close to the circumstances which created emotion — but would not allow it to touch him. The physical manifestation of this was not merely the affectless pose: he did not touch people and he did not allow them to touch him; he did not hold his children or allow them to hold him; he did not kiss them, shake their hands, put his arms around them.

His dress was part of this. Everything was starched and perfectly matched. His handmade suits were protective decoration, too fine to be soiled by the exchanges between parent and child.

The people with whom he bowled, his teammates, all knew him from the factory and understood the type of man he was. They would celebrate next to him but would never attempt to draw him in.

Where my father ran into problems was at events that included people he did not know. There he had less control, and for that reason he stayed away. Activities sponsored by the schools his sons attended — sporting events, dances, theater, chess competitions, these sorts of things — were a source of such annoyance that he could not bring himself to attend them.

On our way home after one of our outings at the Ranch Bowl, Father told me that he had something to say to me. There was a tone here that I had not heard before, a tone that suggested he was about to take me into his confidence.

"Michael," he said, "as you go through life eating Reubens, there will be people who will tell you that it was invented in New York. Don't believe them. It came from Omaha."

The way that this happened, according to the story that Father told me, was that in the 1920s a group of men, including one or two of my great-uncles, would gather in a suite at the Blackstone Hotel to play cards. The owner of the hotel would send menus up to the men, but, bored with the menu, some of the men began to experiment. One of those men was Reuben Kulakovsky, known to his friends as Reuben K.

The controversy arises from those who believe, understandably, that the Reuben sandwich was born out of the kitchen of Reuben's Restaurant in New York. There has been much back-and-forth on this subject, with the proponents of the conflicting schools bolstering their claims by uncovering ever earlier references to the sandwich, as excavated from reviews, letters, and diaries of the time. Both sides have presented plausible evidence that the sandwich dates to the mid-1930s, but with each side stuck at that period, the debate is close to a draw.

I mention this in part because following my discussion with Jack Kawa, the owner of Johnny's Café, where my father would have breakfast after his nights at the Miller Hotel, I noticed a menu pinned to the wall of Johnny's. On that menu was a "Reuben sandwich," and while Jack Kawa could not date the menu precisely, the prices suggested that it was from the 1920s or, at the latest, the early '30s. This gave me some bit of satisfaction, for though I had yet to find the woman in the painting or my black half brothers and half sisters, had learned little or nothing about my father, and had gained nothing like the sort of insight into myself that people achieve on such investigations, I could find comfort in having made a contribution to the history of the Reuben sandwich.

In addition to bowling, my father skied. He would head down the hills of Colorado, Utah, and Idaho, it mattered not where, in the frozen isolation that those places provide. He did not care how many times he fell, sprained his ankle, or dislocated his shoulder; he never complained—skiing kept him away from other people, and these injuries were the acceptable payment.

Once he started skiing, there was no question of a vacation to any place other than one which had a mountain. My elder brothers had no strong affection for the sport, so they would stay home or go elsewhere. Because I was young, I didn't have that choice and I learned to ski at a young age.

The resort where my parents skied was small, and after a few years we came to know everyone who skied there. Of these, many were well known.

None of these people with whom we skied and dined held the slightest interest for my father. He was nice to them when forced to be in their company, but he did not seek them out.

It was our second year in this place when we discovered a restaurant that specialized in Austrian delicacies. In the corner of the restaurant was a man who played the zither. The restaurant was decorated in a style that suited the music and food, and the waitresses and maitre d' wore the appropriate costumes.

Having a late lunch in the restaurant, we happened to notice a man with his family. He had a wife and two children. During a break in the music, the man at the table motioned for the manager and asked him if the zither player knew anything from the musical *Zorba*.

The manager told the man that the musician played only Austrian tunes. Hearing this, the man brought out from under his table a large tape recorder. From the recorder came the music of *Zorba*.

Two days later we were having supper in the dining room

of the hotel in which we were staying. An orchestra from the East Coast had been brought in to entertain, and most of the hotel's guests, formally dressed, danced between the courses. In between sets, we once again heard songs from *Zorba*. There in the corner was the man with his tape recorder and family.

It came to pass that my parents met this other couple, and within days they were friends. Within a year or two, best friends.

The man's name was Oscar, and his wife was Isodora. They lived with their two children in Philadelphia, though that had not always been the case. I became friendly with the son, and it was he who told me the story of his father.

As a young man, Oscar had been a low-level employee in a large tobacco company. Impressed with him, the tobacco company transferred Oscar to Cuba, where he became manager of one of the company's cigar factories. Oscar proved to be an adept manager and his stay in Cuba lengthened. As it did, Oscar began to venture deeper into Cuban society. Where Oscar ended up was in the countryside, in the villages, where he would spend long evenings drinking, talking, and distributing cigars.

This is how it came to pass that Oscar met and befriended those who were planning the overthrow of Batista y Zaldívar.

The time came for the revolution, and the revolutionaries warned Oscar that he should liquidate his company's factories in Cuba and start up elsewhere. Oscar, who was appreciative of the advice, immediately wired his superiors. Oscar's warnings were ignored, but he persisted. In response, the company decided to fire him. Before that happened, the revolution broke out.

With Oscar's friends in power, he felt no real need to leave the country, so he remained there with his wife, Isodora, and son.

Years later Oscar returned to Philadelphia. His friends in Cuba gave him a tugboat and undisturbed passage out of the country. Oscar loaded the tugboat with his family and enough of the finest Cuban cigars so that he could smoke two a day for the rest of his life.

Oscar was one of the few people whom my father genuinely liked, and as a result we spent time with Oscar and his family. Oscar and my father would meet at the end of the day, have a few drinks, and Oscar would start to tell stories or expound on one subject or another. One of Oscar's theories was that the best indicator of a country's economic condition was the price of shrimp; I have no idea what support he had for this, but he would go on about it for hours, explaining the complexities of this to my father, accompanying his explanations with graphs which he would draw on the sides of grocery bags. Father would listen and every few minutes ask a question or add his own observation, but mostly he would sit quietly enjoying the agile peculiarity of Oscar.

I had somewhat the same relationship with Oscar's son, Edward. Edward was more interested in chain-smoking than skiing, and for most of the day that is what he did: Edward would walk around the town, a coffee or drink in one hand and cigarette in the other, reflecting on mathematical puzzles — Edward had an advanced mind and set it to predicting, through elaborate and to other minds incomprehensible systems, those phenomena that were assumed to be random. In this way, his mind was much more specific than his father's.

What distinguished Edward from the others in town was that he looked Cuban. Neither Isodora nor Oscar were Cuban but Edward had the blackest hair and darkest skin that I'd ever seen on someone who was not black or Hispanic. The impression that Edward was Hispanic was enhanced by the way he danced: Edward had learned to dance in Cuba and at a very young age had won several dance competitions. He and I would go to a dance in the resort, and while everyone else was jumping around thrusting their arms about, Edward did sly, sinuous dancing, dancing that moved like the smoke from the mouth of a cabaret *cantante*. After he danced, he would sit at the bar and scribble equations.

During one of our walks, Edward and I came across a room of men playing poker. I'd been introduced to one of these men before; he was the owner of a large corporation. He and the others invited us to watch their game, and as the game progressed (with stakes that I'd never seen before), a look came across Edward's face and the next thing I knew he had invited himself into the game.

Edward's clothes were ripped, he smelled of drink and cigarette smoke, and though he was not allowed to carry a lot of money with him, he had enough on him to get himself into the game. The men saw Edward as someone who could be easily taken, and in this they were right. Within an hour, Edward was out his money. This did not bother me as much as the smirking of the men when my friend Edward was forced to leave the table.

The next day Edward announced that he had been up all night developing a system to win at poker and that he was ready to return to the game. That afternoon we walked back into the room where the men were playing. Pleased to see us, they offered us a place at the table. I declined, but volunteered to deal.

Edward lost a few hands and was on the verge of another humiliation—a humiliation deepened by Edward's childlike confidence in his system—when matters turned. Quickly Edward won his money back and more. By the end of the game, Edward had prevailed. Edward was generous enough to share his system with the others, who scribbled down the equations. Whether they ever used them, I do not know; but I've wondered.

That evening, my father told me that he'd heard that Edward had won a good deal of money at a card game, and he wanted to know if I had anything to do with it. I told him that I did nothing more than deal the cards. He asked me whether I still had the cards. I did and handed them to him. He turned the deck over.

The human eye is such that when separate images are presented to the eye at a rapid speed, the brain perceives the illusion of continuous motion, a phenomenon first described by Roget in his paper to the Royal Society of London in 1824 — "The Persistence of Vision with Regard to Moving Objects." This work, which provided the theoretical basis for the movies, has an application for card players: because each card of a marked deck has a different design on its backside (the design which tells the dealer what card is being laid down), if one stares at the backside of the deck and at the same time flips quickly through the cards, there will be the illusion of continuous motion, a sort of dance.

Father tossed the cards into the fireplace.

Edward and I remained friends, and over the years he and I would get together whenever we could. Edward would change many times in those years — his ideas, mannerisms, dress. One day I was told that Edward had been institutionalized.

So insistent was I in finding continuity in him, in discovering a persistent essence, that I ignored what he had become. In the subsequent years, I have wondered whether in flipping so quickly through our lives and the lives of others, we have created the appearance of continuous selves where there is none.

4.

WHEN I REACHED my junior year in high school, I inherited a car from my father. That car was old and extremely large—one of the largest cars ever made, and completely unlike what anyone else in the school was driving. It was bright bronze on the outside, beige inside, and had soaring fins and a convertible top.

There was an outdoor parking lot around the corner from the school, and the man who ran the lot gave me a space for the car. I think he felt sorry for me. He had a small booth at the edge of the lot and assured me that the car would be secure.

The car allowed me to explore areas of the city that I had not known.

Last call at the bars in Omaha was 1 A.M.; last call in Council Bluffs, Iowa, which was directly across the Missouri River, was an hour later. With the car, I could drink in Omaha till one in the morning and then get to Council Bluffs for another two or three drinks. James Ross III and I would take turns driving to Council Bluffs. James also had an old convertible, though his was red.

It was common for the police to stop kids who were in high school and search their cars for bottles of alcohol. This proved to be somewhat of an inconvenience, until James, who enjoyed having a Scotch while he was driving, came up with a solution:

He emptied the container in his car that held the windshield fluid. Then he filled that container with Scotch. The last step was redirecting the tubes from the container into his glove compartment. The result was that he could fill the crystal glasses that were stacked in the glove compartment by pushing the Wash button on his dashboard.

At the time, downtown Council Bluffs was full of strip bars. They were pretty relaxed about whom they let in, so James and I went to these places.

Other than the strippers in Council Bluffs, my experience with women was limited. The first girl in whom I had a real interest was the young woman who sat near me in the class where I had met James—this girl, people thought, was Native American. This led me to do quite a bit of research on Native Americans so that I would be prepared when I was called to the reservation to participate in ceremonies with her family. As it turned out, she was not Native American.

Later that year she and I were paired up in a chemistry class, which meant at least one night a week I was at her home working on an assignment. Before the night was out, each member of her family would stop in to wish her sweet dreams.

One evening, after the others had gone to bed, she remarked that she had something to show me. As we reached the third floor, she took my hand. There were two rooms on the floor, hers and a bathroom.

She brought me into her room and sat me down on one of the beds. Still standing, she turned around so that she faced the other bed. Grabbing the edge of the bed, she bent over slowly, her skirt rising to the top of her thighs.

As I reached out to touch her, she pulled a trunk from under the bed.

Her fingers floated across the lock. Within seconds, the lid was open: inside was a vast accumulation of confections.

Her feasting lasted for an hour. She was curled over the trunk, the wrappers climbing her thighs, her mouth syrupy, her teeth licorice-black. Finally she stood and left the room.

When she returned, her blouse was off; her breasts were

glazed with sweat. As her eyes cooled, the fragrance of her in-
sides held me against her.

My younger brother, Bruce, was having better luck with
women.

On a trip, we had run into the Pritzker family from Chicago. They owned the Hyatt hotels. My mother knew the father
of the family from college, where they had dated. They remained fond of each other.

This fellow was skiing with his family, and one of his kids, a
daughter, was a year younger than my brother Bruce. Bruce
and the girl got along well. My brother liked her but the two
were very young.

When we returned to Omaha, we recounted the events of
our vacation, including the happy reunion of my mother and
her old friend from college, and the story of Bruce and the
friend's daughter. A sour look came over Harlan's face. To him
the story of Bruce and the young lady was reminiscent of an
earlier story (her father's unrequited affection for my mother),
and this opened the top of a more general truth: my siblings
and I were the product of at least three families that shared a
remarkable talent: the ability to dissipate large sums of money
while simultaneously running from all opportunity to replen-
ish those fortunes.

Before we were done with lunch, Harlan had left the table.
We heard him in the other room talking on the phone.

Shortly he returned.

"We leave in an hour."

He then turned to me: "Ready the car!"

Minutes later I was at the wheel of the car, Bruce and Harlan in the back, and next to me Grandmother.

Under Harlan's instruction, I headed down Dodge Street.
He had not revealed where it was that we were going, but the

day was a fine one, the top was down, and after a large lunch and some drink the drive was refreshing. The further downtown we drove, the fewer the options. At Abbott Drive, I picked up speed. We were on our way to the airport.

Harlan's plan was then disclosed. Bruce and he were leaving for Chicago, where they would meet with the young woman and her father. Harlan would make clear Bruce's true feelings, which I suspect were still hidden from Bruce; whereupon Bruce and the daughter would be excused while Harlan and the father made arrangements for the future joining of the two families. A bold but necessary plan.

"Michael," I heard Harlan shouting from the backseat. "You will have the Hyatt in Puerto Rico."

We were passing the last motel before the airport.

"Bruce," Harlan continued, his voice filling with excitement, "you will run the Hyatt in Chicago, the flagship."

I noticed a smile cross the face of my grandmother. The Taxmans and the ancient house of Rips would be returned to their proper stations—stations gained and lost generations before the Pritzkers had heard of miniature shampoos.

Harlan was not through.

"I," he exclaimed, his voice barely audible through the wind of the speeding car, "I will have the Hyatt in New York."

In the rearview mirror I could see Omaha in the distance—the small buildings, the flatness, the niceness. We were leaving it all behind.

"And Lance," the voice boomed from the back, "the Hyatt in Berlin."

Berlin?

I looked into the rearview mirror. All six feet four inches of my brother was standing in the speeding car, fearless against the wind, posture perfect, his right arm extended.

"Hyatt Hitler!"

5.

B Y MY senior year in high school, I rarely ate lunch at the school. On some days my friend Keith and I would go out. On some days I would meet my father. More often than not I would eat alone at one of the old restaurants downtown.

One afternoon I decided to have lunch at the Silver Pit, a small concrete hut that served ribs and pig snouts and which was one of the several places in South Omaha to which my father regularly returned. The cook, owner, and waiter was named Jasper and my father liked him, though the two of them probably said no more than three or four words in the many decades that my father ate there.

Walking to the parking lot, I discovered that my car was gone and the man in the booth was missing. The police caught up with me at the end of the day to say that the car had been found—in the very same spot where I had parked it in the morning.

Two weeks later the story repeated itself: I went to the lot in the middle of the day, the car was gone, I called the police, and they found it in the lot. The third time this happened, I dispensed with calling the police.

What I'd learned from the police was that the car was so old that the key to the ignition was stripped and the car could be started with anything thin enough to be slipped into the keyhole. Someone was breaking into my car, driving it for the day, and then returning it in time for me to get home. The arrangement seemed sensible enough, particularly since every month or so the person who stole my car was good enough to have it washed and filled with gas.

On the weekends, I had the car to myself and would use it to drive into the countryside. Many of these towns continued to be dominated by the same ethnic groups that founded Omaha and South Omaha in the nineteenth century. One of the reasons this had come about is that the United States government had given land to the Union Pacific as an incentive to build rail lines across Nebraska. To raise money, the Union Pacific sold off much of what it had been given, with large tracts going to immigrants, even whole villages of immigrants, who were eager to leave Europe.

Despite my interest in the countryside, it was in downtown Omaha that I felt at home. This too I shared with my father. If he could be convinced to depart from the couch, he would want to go downtown, where he would come across the people he was meeting in his art classes.

A habitué of the Market in those days was a tall, thin fellow who appeared to have little to do but sit in bars. He was not from Omaha, but beyond this his past was opaque. He was, we thought, a decade older than us, but that too was unclear.

My debt to this man—his name was Richard Flamer— arose out of our first conversation. James and I and a friend of ours, Michele, were sitting in a bar when Flamer arrived at the table. Flamer knew Michele and he joined us.

Somehow we got on the subject of writers from Nebraska, and when Michele or James mentioned that we were reading Willa Cather, Flamer asked, "In the English?" which was the sort of remark that I came to expect from Flamer and the sort of remark that distinguished Flamer from anyone I had met before.

Flamer explained that Cather was better read "in the French than in the English," and he had a collection of her first editions in French.

If we were going to read someone in English, Flamer recommended the Nebraskan Weldon Kees. I had no idea who this was, but that was not unexpected. That neither Michele nor James knew Kees was more surprising.

Then and there, Flamer delivered a disquisition on Weldon Kees, including an analysis of his poetry, a review of his careers as a literary critic, art critic, jazz pianist, and early abstract expressionist, and finally his disappearance or suicide (there was debate on this, with some believing that Kees had run away with a young woman to Mexico). The day after Flamer's lecture on Kees, I stopped by the library to see if they had any of his works. After the library closed, I headed over to the Congress.

To my surprise, the Congress was transformed. Not by Eckert or the Colonel or the residents, but by Flamer.

How had he done it?

I will attempt to explain:

As I sat in the lobby that day, I recalled what Flamer had said about Kees—that Kees might well be alive and living under an assumed name. Was it not possible, even likely, that Kees, after years in Mexico, had returned to Nebraska, his native state? And if he was back in Nebraska and wanted anonymity, what better place than the Congress? This also explained the appearance of Flamer in Omaha: he was a Kees hunter.

Among the men at the Congress, whom everyone believed to be down and out, there was one of the most important, though forgotten, authors of the century. All I had to do was figure out which one he was. With this, my trips to the Congress increased.

Not wanting to scare Kees away, I would ask subtle questions of the old men, the answers to which might reveal his identity.

My growing obsession with finding Kees was having an unfortunate effect on other parts of my life. About halfway through the semester, I was informed by my French instructor that unless my grades improved, I would fail the course. Part of this was Michele's fault: she was superb at foreign languages and could be counted on to get me through my homework; but that year she was in another French class and was too involved in a variety of activities to save me from my ineptitude.

But if Flamer had led me into this mess, he also led me out.

The high schools of Nebraska had decided that they should do something to encourage the study of foreign languages. This was a more complicated project than one would assume: studies had shown that many of the students who proved reluctant to learn a second language actually knew a second language. This paradox was explained by the fact that these students were the sons and daughters of immigrants and, consciously or unconsciously, were embarrassed by their parents and the foreign languages they spoke.

To rectify this, the high schools announced a statewide competition based on an original composition written in a foreign language, or an original translation of a preexisting work. The winner would receive a substantial prize. Students who entered the competition were not obligated to finish their foreign language classes; their compositions or translations would be graded and that would stand as the grade for the course.

Until my conversation with Flamer, I had given no thought to the foreign language competition. But he had said something about Cather—"better in the French than in the English"—that inspired me.

Entering the public library, I approached one of the librarians. What I was looking for, I told her, was a translation into French of a relatively short work.

She told me that nothing came to mind but that I should come back in a few days. I thanked her.

A few days later I returned. The librarian appeared with an armful of books.

I asked her what she knew about the authors.

There was only one author about whom she knew nothing. That was my book—an American author so obscure that no one would suspect that I had found a French translation of his work.

I spent three evenings retyping the French text onto my own paper, prepared a cover page, affixed it to a bright red folder, and presented it to my teacher. She could not have been more pleased—the foreign language competition had inspired a miserable student to do something exceptional.

How exceptional, I did not know until I happened to brag to one of my older brothers that while everyone considered me a lousy student I had succeeded in translating an unknown writer named Ionesco into French. The look on my brother's face was something like what I'd imagined the expression was on the faces of those who actually saw the devil peeking out of a woman's pubic hair, which is to say that unlike the librarian, he knew who Ionesco was and that Ionesco wrote in French. My bumbling, he assured me, would soon be exposed and I would be expelled from school.

I needed to get myself out of the competition. But it was too late. The French teacher had already informed me that my submission was by far the finest that she had received, and though it needed corrections, she was putting it up against the entries submitted by the other language teachers.

The critical moment came two months later, when the principal announced the winner of our high school competition: an original translation into French of Ionesco's *Rhinoceros*.

By winning the high school competition, my "translation" would now be up against the winners from other high schools across the state, and that competition was to be judged by a local university. What's more, I would be obligated, as would others in the competition, to read the translation to the judges. Members of the public would be invited to the readings.

Walking out of the auditorium, I was stopped by my French teacher, who told me that she had decided that *Rhinoceros* should be performed by actors, instead of read, and that she knew a gentleman who she thought might help.

Two days later I met that fellow—a short man with a pocket square.

"Since you are not a trained actor," he began, "I will not ask you to act, but I insist that you sit with me at rehearsals. Do not forget, this is your translation, and it should be true to what you want."

And that is how I spent my next few months—sitting next to him as he guided his student actors up and over the hurdles of absurdist theater. Michele assisted with the pronunciation of the French.

As the competition drew near, the director and I were convinced that the play we had created—a play that neither one of us understood—was nothing short of extraordinary. And on a lovely spring day, we gathered on a bus and headed to the university.

It would please me if I were able to say that by this time, having resigned myself to the catastrophe that lay moments away, I was able to enjoy myself. But that was not the case.

There were three judges on the panel—each serious, and each, I believe, involved in the arts. To my dismay, they all looked intelligent.

"Here we go," said the director, his voice excited. "They will talk about us for years."

About that there was no doubt.

As the play began, I closed my eyes. The French came and did not stop; a tormenting shower of it.

At one point I glanced toward the judges. The faces of two were stern—but they had started out stern. Even if they suspected that Ionesco wrote in a language other than English, it was possible that he wrote this one in English. Wilde had written *Salome* in French.

But the third judge, the one in the middle, knew. He had read Ionesco, knew that Ionesco wrote in French, that Ionesco had written *Rhinoceros* in French. And yet he was still not certain. Yes, he was certain of Ionesco; but he was not certain of what he saw before him. The fluctuating expressions on his face suggested that he was debating between one of two explanations for what he was watching: a daring parody (a send-up of high school academics, foreign language instruction, and theater itself—a mocking of absurdist theater through the vehicle of one of its greatest works) or, something equally remarkable, a highly competent performance of absurdist theater by a group of imbeciles—misled by comical accident into believing that they were performing a translation.

After the performance, the director and actors went to lunch.

At three in the afternoon, everyone gathered to hear the results. I slipped into the back of the auditorium, thinking that this would give me a few minutes before the marshals caught me racing across the campus. I saw the group from my school sitting toward the front.

One of the officials of the university was standing on the stage. He spoke of how honored he was to have hosted the competition and that he was certain this would encourage the study of foreign languages. He pulled a piece of paper from his jacket and read out the finalists.

I was one of them.

Before I had a chance to get out of my seat, I heard something that induced paralysis of my voluntary muscles (what the medical texts refer to as Locked-In Syndrome, and for which there is no known cure). The speaker had announced that the university was so proud of this event that the university press had agreed to publish the winning composition.

Up to this point, whatever offense I had committed was one of a moral nature—something between me, my family, and my school—but now I was close to committing a felony and bringing the university down with me. I had to get out of my seat.

"Help me up," I said to the elderly woman next to me. "I'm paralyzed. I have Locked-In Syndrome."

She got up and tugged on my arm. But it was not enough.

"Harder," I implored.

The tugging caused several people to turn around; one of them the middle judge. He smiled; and at that moment I knew I was saved.

Whether we were parodists or idiots, he thought the whole thing so rare—so much something that Ionesco himself would have appreciated—that he could not resist doing his part: he had made us finalists but not winners. As the winners were read, he came over and lifted me out of my seat.

6.

DURING MY YEARS in high school, with Lance and Harlan away, I spent increasing amounts of time with my father. Not that we spoke. He would read the newspaper, I would scribble notes to myself, time would pass. Once or twice he would look up and ask me a question ("The Mauro Castle, does it mean anything to you?") or recite something from memory ("'What age so large a crop of vices bore? Or when was avarice extended more?'").

By this time I was aware enough to appreciate the oddity of

this man. While there were other fathers who had little interest in their children, they made up for it with an interest in themselves.

My father had no interest in his children and none in himself. The respect that other men sought among their peers, the standing that other men gained through philanthropic or religious involvement, were of no concern to him.

Those with greater knowledge of the human mind than myself might suggest that this man was depressed, alienated from his life, potentially self-destructive. But according to every member of my family, every person from his factory, every friend, Nick Rips was content. He laughed, sang, and danced; for him, life was an enjoyable, small place.

And I enjoyed being in his company, as much as one could be in the company of a man who could not be seen. And he, I believe, appreciated the fact that I made no demands of him. He would bring me along to various appointments, and I would remain quiet, record his habits, and learn from them.

His clothier, for example.

Father would regularly stop in to be measured for a suit or topcoat, and I would sit by and watch as he and Sol Parsow, the owner of the men's shop, would discuss the merits of fabrics. Sol, who was from New York, came to Omaha as a soldier and fell in love with a woman from Omaha. The two of them married, and Sol started his own business.

Sol, gregarious and affectionate, was the opposite of my father, but the two got along well. My father appreciated Sol's knowledge of fabrics and tailoring, and there were few men my father listened to with quite the attention as Sol. In addition to clothes, Sol had an interest in a number of other subjects, for as he fitted my father, Sol would discuss lacquer boxes and walking sticks. There were guns on the walls, and from this I developed an interest in shotguns and the engraving of shot-guns.

Though Sol sold shoes, my father bought his shoes in the factory next door to his office. It was my father's luck that in downtown Omaha was one of the world's great boot makers, Dehner's. As a child he had lived across from the family that owned the company.

After a day at work, my father would take me to Dehner's, where we would chat with the owners and inspect boots. Dehner's specialty was English riding boots, and every boot was made to order and every boot made by hand. Dehner's may have had a couple of customers in Omaha, but most of their boots were shipped elsewhere.

A decade or two ago, at the request of someone in the Omaha police department, Dehner's designed a motorcycle boot. My father loved that boot. Between Dehner's, Sol's, and trips to clothiers abroad, there was never a moment when my father was not perfectly dressed. Which is why I thought it odd that he would appear at my high school in his pajamas.

The events leading up to this incident had begun several days before, when I was expelled from school. The terms of my dismissal were such that I would be out for a period of time, at the end of which I would return to school but only if accompanied by one of my parents; we would meet with a vice-principal, the infraction would be reviewed, contrition demonstrated, and a decision made as to whether I would be allowed to return.

Neither of my parents knew of the incident, so after a few days at the Congress, the time came when I had to speak to one or the other. It was not an easy choice. Unlike her husband, Mother cared a great deal about how we did in school, and my transgression would certainly upset her. Father, on the other hand, might not be upset, but if he were, his reaction was likely to be extreme.

Not long before, my brother Bruce, Mother, and I had been riding in the car when Father cut off a truck. At the next light the man pulled his truck over and got out of the car. He yelled an obscenity and taunted my father to get out of the car. The driver was young and muscular. He was also holding a wrench.

Instead of driving on, Father opened the door.

"Where are you going?" Mother asked, though she knew the answer.

"To kill him," were his words—devoid of excitement.

What the driver saw—an expressionless middle-aged man walking toward him—the driver did not like, and he got back in his truck and drove away.

When my father returned to the car, he seemed disappointed.

Then there was the night that he tossed a cooked bird at my brother.

There was also the gun in his desk.

There was also the "turmoil" that had caused the ulcer.

Despite this, I decided to take my chances.

Mother had gone upstairs, Bruce was in his room, and Father was lying on the couch on the first floor. Entering the room, I paced back and forth next to him. He watched me for a few seconds, and as he did, something occurred to him.

"Did you realize," he pointed out, "that if you had a broom in your rectum, you would be sweeping the floor?"

I had not thought of it.

His observation had ended the silence, and from there I told him what had happened:

I'd been sitting in a study hall reading a magazine, half listening to the teacher's lecture on the importance of maintaining absolute quiet, when the student next to me, who was staring with unusual energy at the teacher, asked me what it was exactly that the teacher was saying. The student had prepared for the study hall by consuming hallucinogens and was now having trouble appreciating the full meaning of the teacher's talk.

I told the student that the teacher, who was holding a pointer that could have been confused for a crop, had just announced that he'd decided to teach us all how to ride and that the horses would be coming through the study hall in a few minutes. With that I returned to my magazine.

I was not very far into my article when my reading was interrupted by excited whispering. As I looked up, I was surprised to see the student with whom I had just been chatting standing on the stage in a semisquatting position. If one did not know that his horse was coming, one would have thought he was about to evacuate. From the look on the teacher's face, I knew that I would not be seeing the student for some time.

That, it seemed to me, was unfair. At the very least he should be allowed his opportunity to ride a horse.

I got up out of my seat and walked toward the front of the hall. Stopping below the stage where the teacher and student were standing, I turned around and made the noise of a horse.

The student jumped on. After getting the hang of it, he was able to ride with only one hand on my shoulder, freeing the other to swing in the air while I tossed out the odd whinny. The student even shouted out a direction or two before horse and rider were escorted from the hall by security guards and the teacher. As I descended four flights to the principal's office — the rider still atop my shoulders — I wove back and forth under his weight but did not have the heart to ask him to dismount.

When I'd finished with the story, my father, who had listened without saying a word, went back to his book.

Early the next morning, he stood over my bed.

"Let's go."

With that, I rose, dressed, and walked out to the car. He was there, sitting behind the wheel, still wearing his pajamas.

At school, we entered the building on the side farthest from the office of the vice-principal and made our way, me following him, in the most indirect route possible, to our destination. Teachers and students were impressed with the sight.

At the vice-principal's office, we took our seats in a reception area. The vice-principal's secretary knew me and reported our arrival.

The vice-principal was not prepared for what he saw as he opened the door: a student, in a sports coat and tie, and his father, in a pair of powder-blue pajamas and slippers. Staring at my father as we took our seats in his office, the vice-principal explained what had happened, the seriousness of what had happened, and the possibility that I would be permanently expelled.

But there was now no possibility of being expelled: it would have been irresponsible for the vice-principal to return me to the custody of a man sitting in his pajamas.

Was this madness? I do not know. Perhaps.

On the other hand, I had belittled my father, for it was he, not I, who was being called to the vice-principal's office. It was he, not I, who would have to provide an explanation, to plead for readmittance. The order of things had been inverted; and there was no better way to demonstrate this, in my father's mind, than to appear in his pajamas—the Greek for dress (*katastole*) being derived from the robes worn by the heads of state (*stole*), with both taken from *stello*, the verb meaning "to place in order."

That night, I met James for a drink at the Saddle Creek Bar. Though it is said that the Saddle Creek Bar was the home of the foot-long hot dog, among some it is better known for its tamale float. This particular food came about when it was discovered that the thickness of the Saddle Creek's chili was such that a tamale, if inserted vertically into a milk shake container of chili, would remain perpetually upright, its head just above the surface. Once one got over the appearance of the chili, which had the texture and reddish hues of blood sausage, there was the body of the tamale, marinating mutely within.

By the time James arrived, I had finished off two drinks, was halfway through a tamale float, and my encounter with the vice-principal and my father had been forgotten.

James and I drank and talked of graduation. The president of our class, Celeste, a smart and well-liked black woman, would give a speech. She was a friend of mine.

Knowing this, James reported that a couple of weeks before, he was driving down the street and pulled up next to Henry, a classmate whose father owned one of the biggest car dealerships in the city. Henry was large and white, with short blond hair; he was also shy. He was driving one of his father's cars and had the top down.

At this point James interrupted the telling of the story to make a phone call. James's father, a hemophiliac, had been in a car accident. Though not the first time that James's father, having drunk too much, had ended up in the hospital, he was now much closer to death than he had been before.

Returning to the bar, James quickly finished the story.

At the light, Henry and James started to talk. During this exchange, James noticed that there was a lawnmower sitting next to Henry. James asked Henry about it and Henry said that he was going over to Celeste's house. He was in love with Celeste and could not figure out how to tell her of his love other than by mowing her lawn.

James explained that he had to go. We walked to the parking lot, and he climbed into his red convertible. I asked him if was going to the hospital and he said no. As his car jetted from the lot, his head disappeared into the crimson stream.

Because James's father had drunk too much and his blood would not clot, he was lying alongside death in the hospital. Because James's father was lying alongside death in the hospital, James was driving his red car faster than needed and to a place that neither he nor I knew. Because James was driving in that way to that place, I knew that we were all floating on the blood current of our ancestors, sometimes slowly, sometimes quickly, but always perilously, our heads just above the stream.

7.

IT WAS SELDOM that we were called to a school assembly. But on this day we were marched into the auditorium as soon as we arrived at school. The principal had been prevailed upon by a group of parents to allow them to speak to us.

There on the stage were the parents, each of whom presented their views on the racial disturbances that had occurred in the school. The leader of the group was a Jewish woman, who was the mother of a student at the school. She began by taking us back through the experience that her parents and other Jews had suffered in the concentration camps. As a result of this, she told us, she understood the suffering of black people and wanted blacks in the school to know that they were not alone.

All of the speakers assured us that they were doing everything possible to end racism and were not afraid to identify those in the white community who were exploiting minorities. With this she began to read bravely from a list of names.

It was at this point that a black student, whom I knew, began to chuckle.

I wondered about this. His laughter, I thought, might have been a response to the assumption that the speaker understood his suffering. She was upper middle class, had grown up in the United States, and was held in high regard by the community.

Beyond this, the Jews in Omaha were an integral part of the city—one of the most popular mayors of the city was Jewish, town councilmen (including my aunt's father) were Jewish, there would soon be a Jewish senator, officials in the police department were Jewish, and banks extended loans to the first Jews in Omaha and with those loans Jews built businesses. And now this woman was telling the black student sitting at the end of the sixth row that she understood him.

Beyond that was the question of whether one can ever know the mind of anyone else (the "other minds" problem that had bedeviled the likes of Austin and Wittgenstein). And if such minds cannot be known, then the invocation of shared suffering (an example of which I was witnessing onstage) may be nothing more than a device that we use to make ourselves comfortable with circumstances that should always be strange and unsettling.

We suffer for people in order to absorb and thereby neutralize them—not simply their mannerisms, their way of speech, but the most intimate, unspeakable, and hence frightening part of who they are. Levinas, the philosopher, knew this. If my friend at the end of the row had told the woman onstage that he understood her suffering, she would have become resentful, dismissing him (rightly) as incapable of understanding her, certainly Jews in central Europe in the 1930s, certainly Jews in Buchenwald.

In the middle of these thoughts I noticed that teachers and students were staring at me.

My father's name was on the speaker's list.

With the speeches over, I left the auditorium and made my way down a hallway, up a staircase, and then waited near a back door until I thought that no one would see me leave.

"Where are you going?"

I turned around.

It was the black student who had been sitting at the end of the row.

"Nowhere."

He knew it was a lie. I was going to see the man whom others found so odious.

The black student and I walked toward the parking lot.

The student told me that as he was leaving the auditorium the principal had reprimanded him for laughing during the woman's remarks.

I asked why he had laughed.

He said that when the woman announced that she was working on a solution to his suffering, it occurred to him to raise his hand and explain that he couldn't speak for others but that his own suffering was a painful combination of sexual frustration and boredom and that finding a solution for that was probably nothing she would want to undertake. In addition, he could not imagine why anyone would choose to suffer for someone else: do something for another person, perhaps; attempt to understand what was causing them to suffer, of course; but actually choose to bring their suffering into yourself, never.

I did not know if he was telling me the truth, but I appreciated that he was attempting to make me feel better.

Years later I received a letter from him. He was in graduate school. On the letter he had drawn a stick figure of a young man with his hand raised. In a bubble above his head was written: "*'Das Leiden selbst wird durch das Mitleiden ansteckend.'*"* Better just to mow someone's lawn.

At the factory, my father's secretary told me that he had left for lunch; after checking with Helen at the Castle Hotel, I walked down to the Mauro Castle.

Looking through the window, I saw Father sitting alone at a table in front of the bar. The owner was taking his order, and the two were exchanging the words that they had exchanged for decades. Drinks and sandwiches were being pushed across the bar, people filed in and out, sunlight chased shadows around the room, shadows chased sunlight, the bartender told a joke, people laughed, the door squealed, and then I lost him; he had disappeared into the laughter, the shadows, the sunlight, the glass, the plate filled with half-chewed bread and bacon.

The attempt to unionize my father's factory led to an enormous struggle. Father believed that the unions would restrict his ability to get rid of inefficient workers and promote those who were more deserving. The quality of what was produced would suffer, endangering the company. Moreover, while almost all other companies had left downtown Omaha for the suburbs, my father had remained, hiring those who had been left without jobs.

The unions argued, effectively, that the workers would receive increased wages and security if the unions negotiated on their behalf.

Steven Bloch, a prominent attorney and a cousin of my fa-

---

* "'Pity makes suffering contagious.'"

ther's, represented the company at this time. With the vote on unionization approaching, he reviewed with Father the means available for defeating the unions. My father shook his head. There was only one way, my father responded, to secure the vote of his workers: he would speak to each of them personally; it was his company, he hired them, he worked with him, they knew him. If they were going to reject the unions, it would only be because they trusted him. With that, he walked into the factory.

The workers rejected the unions. It was this that led the woman in the auditorium to denounce my father.

Before leaving for college, I went through my papers. I came across several photographs, one of which was of a young girl, whom I believed to be the little girl Gretchen. The girl Skimmy had brought into our house to console my mother.

There were other photographs, including me in a sailor's suit, me with my brothers, and me alone with my father, and with these in hand I walked downstairs to reveal my discoveries. My father was alone. Sitting next to him, I handed him the photographs.

The first one that met his eye was me in a sailor's suit. To this, he responded: "The cabin boy was chipper, pernicious little nipper. He stuffed his ass with broken glass and circumcised the skipper."

If there was ever a chance that I would again be party to such a costume or anything like such a costume, he had snuffed it. As to the other photographs, he showed little or no reaction.

Before leaving him, I recalled that I'd forgotten to ask him about the little girl.

"Gretchen," he said without glancing at the photograph.

And here I had him on an oddity that even he would have difficulty ignoring. Once they had taken her in, how could they have let her go, and how exactly did they dispose of her?

"I am curious about Gretchen," I began.

The book which was covering his face came down. His face, at which I rarely looked, was now visible. There was some of me in that face but there was also some of someone else — Gretchen.

Suddenly it could not have been more obvious: I was Gretchen. I was the charming Fräulein.

Father had passed his looks to me and I had passed them — through the medium of a transvestite experiment performed by an elderly English nanny — to Gretchen.

I was ready to leave for college.

THE TORNADO
AND THE THIRD STORY

I.

As I was standing with my father at the airport, a young woman appeared next to us at the ticket counter. I had met her once before, briefly. She too was on her way to college, and my father thought it would be a good idea for me to sit next to her on the plane. The necessary arrangements were made.

My father was convinced that women, all women, were attracted to him — an understandable assumption, I now understood, for someone raised in a whorehouse. This attitude manifested itself in an easy, seductive way, and this, in turn, made him attractive to women. His sons, he hoped, would follow his example, and he made every effort to surround them with attractive companions.

The woman from Omaha and I were not very far into our flight when she informed me that she had graduated at the top of her class and was on her way to a prestigious college. She was much too clever to say this directly, but the point was made. What she did not have to say, directly or indirectly, was that her hair was long, her eyes large, and her body remindful of a certain circus performer.

Our talk was going quite well when I noticed that this otherwise confident woman was beginning to drop her head, as if suddenly shy.

"Is there something that you would like to say to me?" I asked her.

"Yes," she said, still staring downward.

"It's best then to speak your feelings," I encouraged her.

"Your foot."

The conversation was not going as I'd hoped.

"My foot?"

Not certain how it was that I was unable to understand a sentence that began with "your" and ended a word later with "foot," she continued—this time speaking very slowly:

"Yes, your foot. There is something wrong with your foot."

I looked at my foot.

Two years before, I had been diagnosed with gout (one of the earliest cases that the doctors had seen) and was immediately placed on medicines. That helped, but from time to time an outbreak would occur, as it had the day before I was to leave for college.

Before getting on the plane, I'd taken a scissors to my shoes. This was not enough. The change of pressure in the airplane caused an even greater swelling, and I was forced to slip both feet out of my shoes to relieve the pain. By the time she spotted them, my toes were enormous.

To deflect her attention I told her that my father had a far more repulsive condition—a disease that was only known to certain parts of Africa, a disease that had caused his toes to go green and had we not intervened would have spread up his legs and across his genitals. I had personally participated in the burning of the shoes with Roy, a man who walked down stairs backwards.

She had stopped looking at my feet and was now looking at me: my toes distended, carrying a cane, smelling of alcohol, and now reciting a story about my father that I myself had trouble believing. I was an abomination.

As soon as the plane landed, I went to the bar at the airport. The woman went somewhere else.

Arriving on campus, I immediately conducted a survey of the local liquor stores. There were several with full selections. Ordering what I needed, I proceeded to pay, when the owner of the store asked me for identification. In the thinking that had gone into my choice of college, I'd neglected to look into the matter of the drinking age. As a result, I was stuck in a state that would not allow me to buy alcohol.

A week later, a box arrived. It was full of Scotch. The card read, "Your father."

Other than this, Father showed no interest in my college life. Classes, grades, friends were of no concern to him, and if I had questions, I found others to consult.

Several years later he interjected himself.

At the end of my sophomore year, I was chosen for one of the several eating clubs on campus. The club which I was asked to join admitted no women and few, if any, blacks or Jews.

When I asked one of the members why I had been invited to join, he said that they considered me unusual. The unusual people I knew — the fellows from the Congress; the fellow who had congress with chickens; Flamer, with his collection of French editions of Willa Cather; my relative, who had auctioned off his own underwear — came to mind and I was not among them. I suspected that what the student from the club meant was that I was endearing in the way that people have a soft spot for dogs with three legs.

Before I had made my decision on whether to join the club, I was invited to dine there.

The dinner was held in a wood-paneled room lit by candles set in silver candelabra. There was a restrained camaraderie in the room, and the meal, served by waitresses, was a good deal above average. The wine cellar was one of the finest on the East Coast.

The next day I received a call from my mother and, in the course of our conversation, related the details of the dinner the evening before. Mother, who knew about such things, was familiar with the club and thought it worth joining. We discussed the club for a minute or two. Then she passed the phone to Father. He was silent.

By this time I knew the silences as the connoisseur knows the difference in the reds of Rembrandt versus those of Titian, and this one was the silence that I associated with our one trip together to the high school.

My father would have preferred that I dine alone in a restaurant, which is what I had been doing, rather than join this club. His objection was not to their way of life (his own life was one of great indulgence) nor to their politics (he was and had always been a Republican); what disturbed him was their gathering together, their coalescence, one with another with another. They were close and affectionate and all the same.

I declined the invitation.

My view of my classmates was less harsh. In fact, it wasn't harsh at all. I accepted the sincerity of their claim to be struggling for self-knowledge, their claim that with the strength that they gained from that struggle, they would lead us from alienation and unhappiness, their claim that inside each of them was a self that would guide them. If we kept track of that self, returning to it in times of personal and political difficulty, we would be better people, a better society of people. According to my classmates, we had not only a self but a unique self, and not only a unique self but one that was a source of moral authority.

But as much as I admired them, I could not join them. I did not have the confidence they had in the self: I was not certain that it existed or, if it existed, that it had a meaningful claim, moral or otherwise, to authority over my life.

There are a variety of hypothetical cases for testing assumptions about the self. One of these is to assume that all information regarding one's material and psychological makeup is recorded and then transported to another planet where a machine creates an exact duplicate out of the same type of materials that went into the original. As one commentator points out, some would find this a form of murder, while others a fine way to travel.

There is another, the "fission hypothetical," wherein the brain of an individual is divided into two, with both new brains having all the information that had been stored in the original brain. The two brains are then placed into two new human forms. Do the two have the same identities? Is either the same as the original?

These cases are by no means new. They call to mind, for example, the long-standing debate over the consequences of being eaten by a cannibal. The discussion arises when, walking along one day, you happen upon a cannibal. After a tussle, he or she gets the better of you. The problem arises at the resurrection, for if your body must be joined to the soul that was in it at the time of your death, then either you or the cannibal (into whose body you have been absorbed) cannot possibly return to your original forms.

I suspected that a greater part of my classmates' attachment to the self was a fear of what lay beyond it. For them, the world outside their minds was overwhelming and punishing, filled with structures that could not be changed or, if so, only with great difficulty. Their thinking was the political side of the philosophical dichotomy of subjective and objective, the mind versus the forces exterior to the mind. It was this dichotomy that the phenomenologists sought to dissolve.

Husserl, having bracketed all assumptions that were subject to even the slightest doubt, came to the conclusion that truth lay in the "transcendental ego." This ego, though, was not the images, reflections, and memories that distinguish one person from another but rather the collection of aspects that are common to all consciousness. In the end, Husserl discovered an indisputable truth, but he found it not in the individualized self but in its opposite.

My difficulty with the concept of the self as articulated by my peers was not entirely philosophical. Try as I might, I could find nothing in me that looked remotely like what they were describing as a self or a soul or even an inner structure. There were things in there, but they were not what I, or anyone, would choose to lead them through life.

There was another point at which I departed from my classmates: the critique of the culture of our parents' generation was one with which I found it difficult to agree. For my classmates, the suburbs where we were raised were contemptible places—places that bred personal unhappiness and reactionary political views. And that may well have been true for them, may well have been true as a general matter, but it was not what I saw: the suburb that I saw—with its uninterrupted grass and trees and its quiet, small-town way of thinking—was the place that (as Olmsted had intended) helped heal my father.

When my classmates vowed that they would never live in the suburbs, I thought to myself that they were fortunate: they had no need for the suburbs—they were stable; they were angry with what surrounded them but secure enough in themselves to set out against it. Their parents, I suspected, were not so sound—not at least when they returned from the war, not when they had lost their brother; the tranquillity of which Olmsted wrote was something that appealed to them.

The summer after my sophomore year in college, one of my classmates drove across the country and stopped in Omaha. After spending three or four days at our house, her one observation about Omaha was that she was surprised at how "ordinary" my father was.

An hour after she had driven away, there was a knock on the door. We were expecting no one, so I imagined that it was the woman who had just left; having thought about her remark, she had returned to apologize.

My grandmother opened the door. There was a brief exchange and the door closed. It was obviously not the woman.

An hour later I was in the kitchen having lunch when Grandmother entered. She took a seat across from me. The two of us were engaged in talk when, minutes later, Mother arrived. She had seen a car in the driveway that she didn't recognize.

Do we have a visitor?

I knew of no one.

"Yes, we do," answered my grandmother.

This was unusual: at her age, Grandmother's friends were either dead or too infirm to pay visits.

And who is the visitor? we asked her.

"Edward."

Even more perplexing. Edward, my friend, lived in Philadelphia and my grandmother had never met him.

Where is Edward? we continued.

"He is in my room with his pants off."

Mother and I walked down to my grandmother's room, and
there was Edward with his pants off.

Earlier that same summer, one of Edward's formulas, a way
of predicting the outcome of football games, had paid off: with
his winnings he had purchased a car, which he decided to drive
from Philadelphia to Idaho. Somewhere in Iowa, he was in
pain. He decided to pull into Omaha.

Greeted by my grandmother, Edward explained that his
"balls were burning" and that his scrotum had enlarged to the
point that it interfered with his driving. Grandmother, duly
sympathetic, escorted him into her room, assuring him that
she would know what to do once she had a look at it.

This was not such a good idea.

Earlier that summer I was forced to bed by severe pains in
my abdomen. With the lights off and my pain growing, the
door to my room swung open. It was Grandmother. Clinched
in her right hand was a long rubber tube. At the end of the
tube, which extended to her ankles, was an enema bag.

On that night, at the instant, she was no longer frail. Her
feet spread apart, her shoulders strong against the night, she
began whirling the bag over her head. Twice, thrice, and then
she released it. As the bag, filled with boiling water, sailed toward my paunch, I knew it was over.

The bag smacked dead on its target, and I lost consciousness. My next memory was the emergency room.

It was Edward's good fortune that my grandmother had not
yet decided what to do with his testicles. Having examined
them, she left them waiting while she went for a piece of toast.

At this point, my mother intervened and took Edward to the doctor. By the time they returned to our house, the cause of Edward's ailment was identified: his pants. It was his custom to wear tight leather pants, and these pants, combined with the heat and humidity of a Midwestern summer, caused his testicles to blossom.

2.

FOLLOWING that summer, I returned for my junior year of college, and it was during that year that I was stopped by a classmate who told me that Omaha had been destroyed by a tornado. Several calls to my parents went unanswered, and I turned on the news. A tornado, having touched down in the western part of Omaha, had moved eastward toward the center of the city—a path that would have included our house. The report said that the destruction was overwhelming, including schools, factories, stores, and houses; casualties had not yet been determined. It was one of the worst tornadoes in the history of the United States.

Two days later I was finally able to get through to my family. My grandmother was the first one to give me a report of what had happened:

As the tornado swung toward the house, Father ordered the family into the basement. Once downstairs, everyone gathered in the room directly below the kitchen, and Father closed all the doors. At this very moment, the tornado appeared on the other side of the street, flattened the junior high school, and began to pull light fixtures and furniture out of our house. This was when my grandmother was sucked up the garbage chute.

Father had neglected to close the small door to the chute, which led from the kitchen to the basement. As everyone sat in the basement, Grandmother levitated into the chute. By the time anyone realized what was happening, she was gone.

The tornado passed, and my family searched the yard and the trees but couldn't find her. My family searched the neighbors' yards and trees but couldn't find her.

The assumption of my grandmother had ended in the kitchen, where she was whipped unconscious by pans and cutlery. Regaining consciousness, she was crazy with anger, because she thought my father had purposely put her under the chute.

The tornado, having finished with my grandmother, continued northward. Three people were killed. Hundreds were injured.

Finishing the story, Grandmother told me that an elderly man had an eye sucked out of his head. Days after the tornado, as Grandmother recounted it, the man was still crawling up and down the street looking for his eye.

Grandmother's accusation that Father had purposely placed her under the garbage chute reflected a mounting animosity between them. Father would never say anything, but we knew it was there.

As Grandmother grew older, she became increasingly opinionated. She also became more demanding with respect to her own needs.

She would lecture me on some subject; I would listen politely and then do what I wished. My brothers did the same. It was Father who, having made certain that his sons were given no direction, could not bear the idea of his mother-in-law defiling the vacuum.

Grandmother did not seem to care. When she was not offering her opinion on the condition of our lives, she was issuing small directives. When I was a child she would have me run to the department store to pick up white leather gloves. If the color wasn't right, if the fit wasn't perfect, I would be dispatched back to the store for another pair. I would run back and forth until the gloves were right. To be honest, I did not mind this—I had an affection for her and there was nothing I enjoyed more than sitting in her room with her and listening to her stories.

Another one of my grandmother's peculiarities was that she would not drive. She could drive, she simply would not, preferring to be chauffeured by the family. Father's solution to this was to hire a driver to take her wherever she wanted to go, and I believe for a time we did have a driver, but Grandmother refused to get into the car with anyone but a member of her family. She also insisted on sitting in the front seat.

Many hours of my high school years were spent taking my grandmother from place to place. These excursions were usually confined to the late afternoon, so they caused little interference.

That certain people might have viewed this as peculiar did not occur to me until the end of one evening when my date, sitting with me in the car outside her house, asked me who the woman was in the front seat.

One weekday, Grandmother found me and told me to get the keys to the car. On her instruction, we headed downtown but ended up pulling into a parking lot near the Blackstone Hotel.

We walked for a block or two, passed Kaufman Bakery, and arrived in front of a movie house. This theater, along with the rest of the neighborhood, was near its end, but that night there was some stirring as a small line of customers waited for tickets. Grandmother was the oldest and I was the youngest.

When it was over and we were walking out, I asked her what she thought of the picture. What I really wanted to know —though I found it difficult to ask—was what she thought of the scene in which an older man sodomized a much younger woman.

"Nothing I haven't seen before," she replied merrily.

The summer after the tornado, I paid a visit to Omaha. My parents were away but Grandmother was home, and Harlan, having moved back to Omaha, had taken up residence in the Market.

A woman I knew from college was coming through the Midwest to visit her parents, and we made arrangements to meet in Omaha. She and I had met during our freshman year and had become best friends.

The days before her arrival I spent listening to records from my father's collection. That collection was given to my father as a gift from one of his tenants.

My father had surplus space in his factory, which he would lease to other businesses. Every so often a tenant would be unable to pay the rent. If Father did not like the tenant, he would ask the tenant to leave. Under no condition, however large the unpaid rent, would Father sue the tenant. Such a suit would publicly embarrass the tenant, and that was something my father would not do.

One tenant, who was having financial difficulties and could not make his payments, asked my father if he would accept records in lieu of the rent. My father accepted the records and moved them to our house. I discovered them as a child and would play them late at night. When I was an adolescent, the musicians on those records preoccupied me.

That summer I was listening to Sister Rosetta. In the middle of one of her songs, my grandmother walked into the room and sat down. After listening to Sister Rosetta for a minute or two, she posed a question.

"Is she black?"

"Yes."

With that, Grandmother left the room. The next time I saw Grandmother was at dinner. She was biting into her toast with such force that a dark mist of crumbs obscured her face. This continued in subsequent meals, until Evelyn, who worked for us and who had become Grandmother's confidante, took me aside to tell me that Grandmother was upset. Not about Sister Rosetta but about my friend Arnel, who was coming to visit.

Evelyn suggested that I inform Grandmother that I had confused Arnel with Sister Rosetta and that she need not worry about a black person coming to the house. I explained to Evelyn that it did not matter whether it was Sister Rosetta or Arnel who showed up—they were both black.

The night before Arnel's arrival, Evelyn and I were watching a movie when we heard a pounding noise. We got up to see where it was coming from and found Grandmother collapsed against a wall; she was cold, gripping the front of her nightgown, and only semiconscious. We pulled her into the living room and laid her on the couch. Her eyes shifted and she began to gasp.

Grandmother was not a well woman—even before this. She'd had numerous operations and was, throughout her life, afflicted with gastrointestinal ailments that required constant monitoring. Though I was never certain what was wrong, she gave the impression of someone who was susceptible to a sudden passing.

That night there was no time for doctors or hospitals. She was on her way out.

I would have to cancel Arnel's visit. There would also be the matter of the funeral, which would be a large one. My mind was going through the people who would need to be called.

"Dan Bohi," Evelyn said.

"Dan Bohi?"

Our next-door neighbor, a fine man, but not necessarily someone who was close to my grandmother.

It then struck me that Dan Bohi was a doctor, albeit a gynecologist. Evelyn was on the phone, dialing his number.

Minutes later, Dr. Dan Bohi came bounding across the lawn in his cowboy boots. At his side was a medical kit. I appreciated the effort, but I knew that Grandmother was dead. I had an unerring sense of these things.

Dan asked Evelyn to leave the room. I followed her. Evelyn went into the kitchen and sat down. I followed her. We sat at the table and said nothing.

Canadian by birth, German by nationality, Evelyn had moved to rural Nebraska, where she worked on a farm with her husband. She survived the Dust Bowl and the Depression only to have her husband die of influenza. He died before they were able to give him the vaccine; he was still young. The farm was too much for her to handle alone, and she moved to Omaha.

A religious upbringing, combined with the tragedy of her husband's death, gave her an appreciation of God's severity. As a result, she maintained a strict code when it came to how she lived her life, and she attempted to pass that on to me and my brothers.

Upon arriving at our house, she looked upon us like those who, for the first time, cast eyes upon the licentious rites of the women of Gebel. After a decade or two, she became used to us, perhaps even finding us a source of secret amusement.

Evelyn would drive my grandmother around. When Grandmother grew old and had difficulty cooking, Evelyn would help her. Slowly, slowly a friendship ensnared them, and one day I noticed that every evening Evelyn would go into my grand-mother's room to sit and talk.

And now Evelyn sat in the kitchen and her one friend in the world was dying. That is why she thought of our next-door neighbor. Evelyn's husband had died before they could get him the vaccine; she was not going to allow that to happen again. She had given years of thought to this evening.

Dan Bohi came into the kitchen, his hair askew, his eyes barely there. I moved to hold Evelyn's hand.

"Gas," he mumbled.

The next day I picked up Arnel at the airport, and by the time I returned to the house there were several calls waiting for me. Grandmother's attempt to convince us that she was dying had failed, and she was now taking more direct measures, phoning relatives to tell them that if my friend was allowed into the house, Grandmother would walk out the front door and throw herself into the street.

The easiest solution was to exchange places with Harlan: he would move in with Grandmother, and Arnel and I would take his apartment. But Harlan was living with someone to whom my grandmother objected, and there was another development: Evelyn, who had not taken a vacation in years, announced that she was leaving for two weeks, which meant that whoever was in the house would have to take care of Grandmother's per-sonal needs.

Her son, Marvin, volunteered.

My uncle moved into our house, I moved into Harlan's, and Harlan moved into a hotel.

Then something happened which had not happened be-fore, and which no one, even under these circumstances, anticipated: my parents cut short their vacation and returned home. The reason they had done so, I suspected, was to relieve my uncle from the burden of having to attend to Grandmother.

That may have been my mother's reason but it was not my father's.

Upon entering the house, this man, who had shown not a scrap of strong emotion in the entire time that I had known him—the man who remained passive in the face of his brother's and mother's deaths—flew open: my grandmother's conduct, he told her, his voice at full volume, was repulsive, and if she expected to spend another day in the house, she would behave properly; otherwise she would not have to walk into the street, for he would pick her up and throw her. My mother remained silent.

3.

ONE NIGHT when Arnel and I were downtown, we ran into Michele and James. Earlier that evening they had dined with Flamer. Over dinner, Flamer, who had never spoken of his past, told them of his life before he came to Omaha, before he started collecting Kees.

What Flamer revealed, according to James, was that he had been the leader of a troop of paramilitary commandos. Governments from around the world would call Flamer and his men when they were unable to solve a problem themselves. Flamer's last mission, according to James, was to a central European country, where the daughter of an American diplomat had been kidnapped. After several years, the American government had exhausted all attempts to find her and called Flamer.

Finding his way into the country, Flamer began looking for

the girl. At first he had no luck. Then he began to hear rumors: a foreign girl, a beautiful girl, was being kept in a part of the city that was under the control of corrupt officials—an area of drugs and gambling. Flamer moved into the area and found her: she had a new identity. Flamer spent the next few months planning the assault.

It was here that James stopped the story. "You must hear the rest from Flamer," James explained.

I did not press James. The opportunity to hear the story from Flamer was around the corner. My brother held a party at the end of every year, and Flamer would be there.

On the day of the party it was snowing and the snow continued into the evening. My brother's apartment was at the top of a long staircase, with large semicircular windows overlooking the center of the Market. From the apartment to the cobblestone street was probably thirty feet.

A couple of hours into the party, the number of people had grown to such a size that it was necessary to open the windows to bring down the temperature. Everyone in Omaha was in that room. Everyone, that is, but Flamer.

James and I were sitting on the windowsill, and having given up on Flamer, I asked James to finish the story.

James was good enough to oblige:

The assault would not be easy. The woman was being kept as a prostitute, working out of a building that was, day and night, full of people. There was only one door in, and it would be impossible for them not to be noticed. They would have to gun down the guards who were stationed inside the building, find the girl, and then get her out before the police arrived.

Flamer's team waited and waited and then, upon Flamer's signal, charged the building. As Flamer had expected, they drew gunfire from every direction.

Flamer and his men, several of whom were wounded, climbed the staircase until they came to the floor where the woman was held.

Bursting into her room, Flamer yelled, "We're Americans!"

The story was interrupted by the arrival of a tray of drinks. Rising from the windowsill, I took a drink for myself and one for James. As I did so, the tray heeled. The woman holding the drinks was staring past me.

James had fallen out the window.

Death had prepared me for a lot, but not for this.

With everyone else, I rushed down the stairs and out the front door. The intersection beneath the apartment was inflamed by the reflections from the snow, and in that cross of light was James—upright and staring upward.

When I was certain that he was okay, I asked him what had happened. He responded, "She looked Flamer right in the eye and said, 'Yes, but you still have to pay.' She might not have wanted to be a whore, but she sure as hell wasn't going back to Georgetown."

James had reenacted the flight of Miss Rietta.

I didn't see much of James after that. He disappeared. Most assumed that he'd left town. This was as expected, for James was always the most worldly of us. We had all thought that he would end up in a penthouse in New York, smoking cigars, calmly directing his life. There was much speculation as to where he was.

Years later I found out that he had never left Omaha. After his father abandoned his mother, she was having difficulty. James moved into her house and cared for her. None of us would have expected that. To earn money to support himself and his mother, he worked at the car rental counter at the airport.

We live at the edge of change but refuse to see it, until something pulls us out the window or sucks us up a chute or wakes us among the whores.

4.

THE FOLLOWING summer, at approximately nine-forty-five in the evening, a police officer was shot in North Omaha. A "help an officer" call went out, and within minutes patrol cars arrived at the site of the gunshot. The neighborhood in which the shooting occurred had been the location of several race riots.

The senior officer on the scene that night was Herb Walker, the security officer at the airport who lived in South Omaha and who had told me about the Miller Hotel.

Walker quickly learned that a family had been taken hostage in their own house and that it was from that house that the shots had been fired at the officer. Through the telephone company, Walker obtained the telephone number of the house and was able to speak to one of the hostages. Walker was informed that the shooter's name was Elza Carr Jr., that he was on the second floor of the house, and that he was armed.

In the meantime a crowd of people had gathered and were shouting obscenities at the police. Walker cleared the neighborhood and encircled the house with policemen.

Walker's first instinct was to get the hostages out of the house. With a cover of tear gas, the police met the hostages at the back of the first floor and took them to safety.

Walker's plan was now straightforward: cut off the water and electricity to the house and wait for Elza Carr to come out. But Walker's plan was rejected. On the orders of the chief of police, policemen were sent through the front door. As the officers approached the house, the front door opened and Elza Carr stepped out and shot Patrolman Paul Nields in the face.

Tear gas was discharged into the house, and after several minutes the interior of the house caught fire. As a result of the fire, the roof collapsed. A few minutes later, Elza Carr came out the front door. The police fired; Carr fell dead on the front steps. Policemen dragged his body into the street.

Herb Walker has a persistent memory from that evening:

"After we had surrounded the house and cleared everyone off the streets, I took up a position in the house next door to where Carr was. This gave us some protection, and we could look right into the second-floor window where he was holed up.

"As I was giving instructions to the officers around the house and trying to get floodlights into place, I noticed a black man standing alone in front of the house. I knew him from the neighborhood; he was a nice guy, a gentle guy, who sold ice cream to the kids, and he was standing there next to his ice cream wagon, a few feet from the house.

"So I order one of the policemen to go out and get the ice cream man out of there, but the policeman says to me, 'I'm sorry, sir, but whoever does that is going to get their head blown off.' 'You're right,' I said, so the two of us start looking around for someone else to get their head blown off.

"About this time, I notice that the ice cream man starts unbuttoning his shirt. He finally gets it unbuttoned, takes it off, and hangs it on the wagon. He then turns around to the house, faces the window where Carr is, and starts pounding his chest and screaming up at Carr, 'You son of a bitch, shoot me. Shoot me, you son of a bitch.'"

5.

AFTER COLLEGE, I moved abroad.

One Christmas, my parents and brother Bruce came to visit me. At a bar in Rome, Father began on the story of Fred and Ted, cousins of my father who lived in Texas. The two were brothers, had grown up together, and, as young men, shared a house. So close were the brothers that when Fred married, Ted remained in the house. Rumor started up that Fred and Ted were sharing the same woman. Following the birth of a baby, some people said it looked like Fred and others said it looked like Ted. The people in the town talked about it for years.

As I listened to the story, it struck me that Ted and Fred—in growing up together, living together as adults, sharing the same woman, and finally raising a child as though it were fathered by both—had performed the sort of identity experiment which philosophers could only imagine. It further occurred to me that if Fred and Ted had one identity, the question of whether the child was Fred's or Ted's became less important.

Even if Fred and Ted retained separate identities, I could not understand why anyone would show such concern about their romantic life. I shared this with my father. To this my father gave the curt, if mystifying, response, "You would if you lived in Sparta."

That reference, I came to learn, was to the story of Cleomenes, a Spartan king, and perhaps the greatest leader of Sparta. The story of Cleomenes is found in Herodotus.

According to Herodotus, Cleomenes' troubles began with a pact he made with the Athenian leader Isagorus. With Isagorus losing control of Athens to the Accursed, Cleomenes was required, under the terms of the pact, to come to his assistance. As Cleomenes entered Athens, the Athenians revolted.

Cleomenes escaped Athens, but the insult was so great that he vowed to return immediately and conquer the city. Assembling an army, he marched on Athens. Before the expedition reached Athens, the Corinthians (allies of the Spartans) pulled out, and Demartus, a Spartan king, also withdrew. Cleomenes was forced to abandon the campaign.

Cleomenes' frustration with Demartus was such that Cleomenes set out to destroy him. He did this by announcing that Demartus was not the true son of the king whom he claimed as his father (which, I assume, is where Ted and Fred came in) and therefore was not entitled to rule Sparta. Demartus raised the matter with his mother, who assured him that he was the son of the king, or if not, a ghost.

The Spartans decided to settle the matter by consulting the oracle at Delphi. The oracle supported Cleomenes, and Demartus was forced from Sparta.

What the Spartans did not know was that Cleomenes had bribed the oracle — a nearly unheard-of act. As soon as this came to light, Cleomenes was exiled. Cleomenes was ultimately allowed back into Sparta, but he showed signs of madness — striking Spartans in the face with his scepter — and was imprisoned.

Having convinced a guard to give him a knife, Cleomenes began to mutilate himself, beginning with his legs, and when he was done with these, moving on to his thighs, hips, and genitals; still alive, he sunk the knife into his belly. The Greeks, according to Herodotus, attributed the fate of Cleomenes to his decision to bribe the oracle to make false statements against Demartus.

Before Father returned to the States from his visit to Europe, he asked me about my vision. The question was a surprising one. There was no reason for him to believe that I was having difficulties.

On the other hand, he may have been suggesting that I was not seeing him correctly. I had spent many years attempting to follow him into oblivion, making myself into a figure that could exist anywhere, never give offense, know exactly what to say so that people would ignore me, deferring to all semisolid objects, thoughts, personalities—a water which, shed upon the rocks, finds its way to the bottom, awakening nothing. But I may have been mistaken about him; I may have been following not him but his well-dressed ghost.

When he spoke, which was not often, he spoke in stories, allusions, passages from poems—material that Malinowski referred to as phatic communion (language, such as talk about the weather, that is used to promote togetherness but is itself meaningless). Sitting in that bar, I considered, for the first time, the possibility that buried in his stories was a message for me, something substantive, something upon which he thought I should reflect.

Ever since I was a child I had been under the impression that my father had made his way to a place in which man and his consciousness were of no weight—the world of magnificent yet immutable and external forms, the world feared by those for whom man's mind or self or soul was the only certainty. Now I suspected that I was wrong and that he was pointing me, through his stories, in a different direction—toward a resolution of the conflict between the consciousness of man and that which lies beyond it, the conflict between the self and the forces that mock it.

The stories—Saul, Ajax, and Cleomenes—were written in antiquity, involved insanity and suicide (uncommon in the literature of the ancient Greeks and Jews), and centered on men who led their tribes and were essential to their protection. Beyond this, there was little that I could make out.

I returned to Omaha less often.

James still could not be found, Michele had moved away, and Flamer, while mixing a preservative for leather books, had set himself on fire. His skin was charred and there followed a long period of recovery, after which he, too, left the city. People say that he is living in Mexico. Perhaps he has found Kees.

The one person who remained in Omaha and had in no way changed was my father. He did not fear old age, continuing to drink, eat heavy foods, and on occasion smoke. My grandfather had remarried when he was ninety, so my father may have thought death a ways off.

Father continued to work, though there was no financial reason for him to do so. At seven-thirty every morning he was in the factory talking to Fred, making certain that the cutting and polishing of lenses, the system he had spent so many years putting in place, was running efficiently. But he was not able to control everything; his brother, younger by seventeen years, had ideas about the business that my father did not share. Moreover, Father's age meant that the number of years in which he could manage a large factory was limited.

When it became evident that my father and his brother would be better off apart, they decided to divide the business. The plan was that my uncle would keep the main factory while my father would retain a smaller factory and a group of stores.

A few days after my father signed over the factory to my uncle, my uncle and his partner closed it. Fred, Charlie, Norman, the others had gone to work, as some of them had for forty years, and there on the door was a note that they no longer had their jobs. Not a single lens would again come out of that factory. My father had no idea that this was the plan. Nor did my brother.

My mother called me the day the factory closed. In my father's mind, he had betrayed his workers.

To help my father through this, Mother planned long stays away from Omaha. One such trip was to Scandinavia. On this journey, my parents met another couple with whom they became friendly. At the end of the trip, the couple approached my mother to tell her that they thought something was wrong with my father. He seemed depressed, distracted, at times incoherent.

When my parents returned to Omaha, Mother filled their time with events, but all the while she was watching my father. What she noticed was that his sadness and guilt over what had happened to the workers had not abated and that at the same time he was becoming more and more withdrawn. Though he had always been reserved, he could be counted on to enjoy himself. Now he could not manage that.

Mother took him to the doctors for examination. They noted a loss in comprehension and memory and an increasing depression, though they could say little as to what was the cause. One night, lying in bed, my father saw a woman other than my mother enter the bedroom. He slipped downstairs and called the police.

My mother woke to sirens. Looking outside, she found that the police had surrounded the house. Afraid to go outside, she called my brother. He drove over and explained to the police that there was no woman in the house other than my mother.

She shouted at my father. Shouted because she knew that

he was no longer with her, that it was he who was the stranger in the house.

And then she cried.

6.

THE TESTS on my father continued to prove inconclusive. The doctors could do nothing.

He would go in and out of sanity.

I tried to get home as often as possible.

When Sheila and I were in Omaha, Sheila would get up early in the morning to have coffee with Father. She was extremely fond of him, and whatever impairment there had been of his thoughts, it had left unaffected his appreciation of her.

Mother had taped a sheet of paper to the wall of the cabinet where she kept his medicine. On the sheet were instructions for him. He was to mix two tablespoons of medicine into his orange juice.

Father asked Sheila to read the instructions to him. He had not told my mother that he was incapable of making sense of what she had written down. For weeks he had neglected his medicines because he did not wish to disappoint her by admitting that he could not figure out what was on the sheet.

On another trip, I sat in his office and watched as he walked among his employees. He was still handsome, stood straight, unwilling to give in to the swelling disorder. He offered to help them, teach them, but of this he was no longer capable.

Returning from the men's room, he walked into the office where I was sitting and closed the door.

He stood directly in front of me.

"Will you help me with my zipper?"

This man, who had taken such pride in his appearance, had

been in the bathroom for half an hour, attempting to close his pants.

The disintegration of his consciousness continued. Mother did what she could to make him comfortable but the disease was chewing gluttonously.

When he was confined to a hospital, where they fed him through his veins, I received a phone call. Father had attempted to pull the tubes from his arms.

I recalled the time when he got out of the car and walked toward the truck driver who had the wrench. My father, lying in the hospital bed, may have been holding on to one last thought: when he saw the final blackness approaching, across the wreckage of his consciousness, he would pull out the tubes and go at it until one of them was no longer.

7.

WHILE FATHER was dying, a woman asked me whether he had known God. The woman had, in the course of her life, experimented with various forms of faith. She was now old.

My father had never talked to me about God; and I doubt that, even in his most difficult times, he had appealed to God.

Intimate moments with the divine are an extension of the belief that man can experience and understand God, that God can comfort man, that man can translate God's thoughts into rules for everyday life. For Levinas and, I suspect, for my father, the failure of this view, as well as of much of modern philosophy, is that it places man and his consciousness at the center of being. By contrast, Levinas urges that our eyes be directed outward, toward an Other that is never fully comprehensible to man, has no contact with man, and, instead of being an aspect of man's consciousness, ruptures consciousness. At the same time we must do everything possible to resist finding ourselves in God and those around us—in Levinas's words, "reducing the other to the same."

Had I mentioned this to the woman who had approached me about my father, she, quite rightly, would have asked me, what good is a God that is inaccessible? A God so obscure does nothing to improve man.

Levinas's response to this is that it is the very separation of God from man that is the foundation of man and ethics: for Levinas, all of man's encounters with other people—their brutality as well as their suffering—are remindful of the original debt owed the Other (the debt of existence) and call upon man for an ethical response (as a form of repayment). Levinas does not seek to establish universal rules of justice to guide this response—such rules would be a reversion to the totalizing systems that Levinas rejects; rather, he prefers that individuals develop their own responses to the specifics of each encounter.

Ethics is at the heart of Levinas's view of how consciousness and meaning are formed. Instead of being the by-product of man's reflections on the world, ethics is a precondition for those reflections. It is, for Levinas, a type of "optics." In this Levinas may well have been influenced by the prophets of the Bible. But there is as well his reliance on the ancient Greeks: Levinas refers to Socrates' observation, as found in Plato's *Republic,* that the good not only causes us to recognize objects but is even more fundamental in that it is responsible for "their being and reality." It is this that has led some to refer to Levinas as the "Jew-Greek."

It is this aspect of Levinas that explains why the Bearded Priest was attracted to him. One does not, the Bearded Priest is fond of saying, have private moments with God; man's experience of God, the Bearded Priest states, paraphrasing Gregory of Nyssa, is never more than man's experience. The reason for God's inaccessibility is that God wants man to focus on what needs to be done on earth; thus, the closer man draws to God, the more God pushes him away—not toward himself but toward others. Here the Bearded Priest quotes the First Epistle of John: "If someone says, 'I love God,' but hates his brother, he is a liar, for he who does not love his brother whom he has seen, how can he love God, whom he has not seen?"

THE
OTHER WOMAN

I HAD NOT SEEN James for some time and we got together for drinks.

While working at the airport, he had convinced the authorities to allow him to open a bookshop. Given a room off a deserted hallway between terminals, he opened his shop, filling the shelves with books that no one expected to find at the airport. Weldon Kees, for example. He hired prisoners from a work release program and taught them about books. The prisoners would wander the shop, offering exegesis.

So popular did the bookshop become that people, including students, scholars, and anyone else who had an interest in books, would drive to the airport just to sit at the shop. For food, there was a toothless convict who delivered pizza from a stand down the hall. And every once in a while, Herb Walker, the head of airport security, would stop by, throw himself into an armchair, and tell stories about South Omaha and about an ice cream man who took off his shirt to yell at a sniper.

James's father had recently passed away. So too James's brother.

James had also buried Eckert, a.k.a. Dusty Tex, which may have been the hardest funeral of the three. Eckert had been cremated, and James was the closest living relative. At the cemetery, James remembered that Eckert's mother had committed suicide and for this reason she had been buried without a tombstone. Having located the area where Eckert's family was buried, James began to crawl on his hands and knees, trying to find the unmarked burial plot of Eckert's mother. James was on his hands and knees, with Eckert in a shoebox under his arm, when a wind came through the cemetery, lifted the lid off Dusty Tex, and dumped him on top of James.

After having drinks with James, I returned to my mother's house. There was a message from a woman who left no number.

Mary, my mother's assistant and cook, was the only one home. She offered to give me a ride to the airport. I was leaving that day for New York. Mary had known my father and she had been one of his favorites. She, on the other hand, knew that he was a man who needed to be left to himself and she had granted him that.

As we were preparing to leave, I asked Mary if she knew anything about my father's paintings. There was something, she said, that she thought I might want to see. Following Father's death and my mother's decision to move from our house on Sixty-ninth Street, Mary had gone through the house to make certain that nothing of value was left behind. In a closet she had found a painting. Mary had put that painting into a storage locker. She did not know whether the painting was still there.

We made our way to the locker.

I noticed a stiff brown board up against the back wall. As I pulled the board out from the back of the locker, an image appeared—a black man; and then another—a white man. Blood dripped from the white man onto the black man. The painting was signed "N. J. Rips." On the back, in his unmistakable script, my father wrote, "Beatitude IV."

2.

THE WOMAN who had called before finally reached me. She said that someone had given her my name (I suspected that it was Frank Williams, the detective, though she would not say) and that she had a story to tell me. I agreed to meet her.

My journey into my father's life had gone on too long. Fragments of his past had been discovered, and from this I knew more of him than I had when I started. But the woman in the painting was still unknown to me and the stories remained a mystery. I had taken an ordinary man (as my friend from college had observed) and convinced myself that there was something significant to him. I would meet with the woman who had called and that would be the end of it.

The woman I was to meet did not turn around when I came into the bar.

The ice cubes in her glass had nearly vanished and yet she had not yet taken a drink. She did not care about that drink, did not care about that bar, did not care about me.

She was leaving.

Her hands, clutching the lip of the bar, pulled her body forward and off the stool. One foot dropped to the railing under the bar, the other foot to the floor. But there she remained.

She began to speak:

A pretty but shy girl had fallen in love with a man. He was

older and he took care of her. He bought her clothes and took her out. One day she found something in a suitcase in his room. She told him what she had found.

He hit her, and when she said she was leaving him, he hit her again.

She promised him that she would not leave him.

The next day, when he was out, she left.

For weeks she hid in the basement of a friend's house. But one night the lights went off in the basement. She was thrown to the floor and her head was held down on the carpet. She felt the first stabs but when the knife entered her eye, she passed out.

When the bandages came off, she was put on a bus to Omaha, where she had a relative. She stayed in that house for months, afraid to go out, fearing that she had gone mad. Her relative took down the mirrors and pulled the shades so she could not see herself. But her memory was more powerful than any mirror. In her dreams, red wet carpet grew out of her face.

If her mind moved from these images, her memory gathered them up again.

At night she would watch television but that was difficult. She needed glasses.

A week after her eyes were examined, she was taken to the store where she would pick up her glasses. She sat at the counter, her hair covering her scars. As the man lifted the glasses to her face, he slid his fingers into her hair and drew it back. Her face was inches away from his. Her hand rose to cover her face, but it could not get there. His hand held hers to the counter.

Through the lenses, she could see a face. A man's face. He asked her about the glasses and she answered. He adjusted the frames. And when she said they fit, he thanked her.

The story stopped.

The tears that had dropped from her eyes never found their way to the end of her face; they crossed her face and crossed it again, each tear on a different path, until those tears rose to the edges of her scars; and now her face — wetted, smoothed — was the face in the painting.

This man at the optical shop was the first one who had been able to look at her without reacting to her face. So when he asked her whether he could paint her, she agreed. He was polite, paid her for her time, and when he was finished with each of the paintings, he would show it to her. She thought that if this man, this stranger, could see her in this way, without the scars, others could and perhaps one day she could.

This woman's life was full of misfortune, but on that one afternoon — the day when she walked into the small store on the first floor of his factory — it was her luck to have sat down in front of a man who had spent years putting glasses onto the faces of men who had been injured in war. So if there was a person in the city of Omaha to lift those glasses to her face — a man who knew to tuck her hair behind her ears, to hold her hand firmly to the counter — she had found him.

3.

ONE AFTERNOON, something that the woman told me stopped me on the street. It was her description of being attacked — of the repeated slicing of her face.

I had found the key to my father's stories.

Toward the end of the first book of Samuel, when Samuel is called from the dead by the Witch of Endor, he explains to Saul that he has been punished for his earlier refusal, in the face of God's direction, to execute the Amalekites and their animals.

What often troubles those who have read this story is the severity of God's punishment of Saul compared with God's leniency toward David, whose transgressions appear to be of a much greater sort.

Such a reading of the story of Saul misses this: at the time Saul failed to follow God's order, the Israelites were weak and under attack by their enemies. Certain commentators have pointed out that as a result of Saul's failure to kill Agag, the Amalekites continued and later attacked the Israelites in the form of Haman ("son of Hammedatha the Agagite").

Shortly before the end of *Ajax*, the seer reveals to Ajax the reason for his punishment: in the middle of battle with the Trojans, Ajax risked offending the Greeks' greatest champion, Athena, by rejecting her assistance.

Saul and Ajax refused the commands of their gods, but what compounded their crime, made it unforgivable, was that they had done so in a way that endangered the people they were obligated to protect. The story of Cleomenes, in turn, is a perfectly condensed expression of this.

Cleomenes, a king of Sparta, not only ignored the gods but bribed the oracle to misrepresent what the gods were saying. At the same time, he weakened Sparta (thereby inviting attack from its enemies) through his extended and internecine conflict with Demartus. That Cleomenes had turned against his own people is symbolized in his striking his fellow Spartans in the face with his scepter, the Greek symbol of sovereignty. It was this defiling of the face that I recalled when the woman in the bar told me her story.

Cut off from the voices of the gods, having turned against their fellow men, imprisoned in their own thoughts, the three protagonists — Saul, Ajax, Cleomenes — became insane, and their final punishment — the punishment of isolation — was suicide.

These men failed in the way that Levinas, who learned his lessons in the concentration camps of Europe, says that man has failed: we have reduced the Other to the same, lowered the barrier to the transcendent, by assuming that God and those who surround us are identical to ourselves; and as a result we are no longer curious, we no longer learn, we are indifferent toward others. Instead of being engaged in a dialogue with the mysterious and unknown — with those who surround us — we have become preoccupied with finding an answer within us.

The woman's belief that my father had seen her in a way that she had not was both right and wrong — right because he saw something that she did not, wrong because it was not her face that he saw. My father wasn't — as I and the woman had assumed — painting her face; he was painting the face below her face and the face below that and the face below that.

The brothel of his grandmother, the stockyards and slaughterhouses, the community of the Magic City created in my father the fear and fascination of the Other that he carried with him for the rest of his life. It was this experience, along with the awe, silence, and piety that were its gifts, which divided him from others. Likable, kind, respectful but removed — not just from his family and friends but from himself.

Following his brother's death in the war, my father returned to Omaha and found comfort in the lawn and the house. This, though, was not enough, and he began to search for what he had seen as a child. And this is when he started to re-create in his factory, one person at a time, the Other that was once and continued to be unsettling and revelatory. He protected all of that until one day when something happened which was to him unimaginable — his factory was closed, his workers disgorged — and when the blight of that struck my father, his mind grew sad and blind.

I had spent years attempting to find him and had not. He never wanted me crawling around in the grave of his self. He wanted me to find him elsewhere — in the face of the Bearded Priest, in people rising and falling through space, in the woman who was stabbed in the face.

In the face of the woman who now set a coffee before me.

Acknowledgments

Steven Bloch, Karen Catalanotti, Greg Fiechter, Julie Johnson, Romain Lamaze, Jordan Miller, Barbara Rips, Bruce Rips, Harlan Rips, Jane Rips, Lance Rips, Nancy Singer, and Frank Wright. Also, and especially, Nicole Aragi, Anton Mueller, Janet Silver, Lori Glazer, Gracie Doyle, and Bridget Marmion.